Hands-On SAS for Data Analysis

A practical guide to performing effective queries, data visualization, and reporting techniques

Harish Gulati

BIRMINGHAM - MUMBAI

Hands-On SAS For Data Analysis

Commissioning Editor: Amey Varangaonkar
Acquisition Editor: Yogesh Deokar
Content Development Editor: Athikho Sapuni Rishana
Senior Editor: Sofi Rogers
Technical Editor: Dinesh Chaudhary
Copy Editor: Safis Editing
Project Coordinator: Kirti Pisat
Proofreader: Safis Editing
Indexer: Pratik Shirodkar
Production Designer: Shraddha Falebhai

First published: September 2019

Production reference: 1260919

Published by Packt Publishing Ltd.
Livery Place
35 Livery Street
Birmingham
B3 2PB, UK.

ISBN 978-1-78883-982-2

www.packt.com

Packt.com

Subscribe to our online digital library for full access to over 7,000 books and videos, as well as industry leading tools to help you plan your personal development and advance your career. For more information, please visit our website.

Why subscribe?

- Spend less time learning and more time coding with practical eBooks and Videos from over 4,000 industry professionals

- Improve your learning with Skill Plans built especially for you

- Get a free eBook or video every month

- Fully searchable for easy access to vital information

- Copy and paste, print, and bookmark content

Did you know that Packt offers eBook versions of every book published, with PDF and ePub files available? You can upgrade to the eBook version at www.packt.com and as a print book customer, you are entitled to a discount on the eBook copy. Get in touch with us at customercare@packtpub.com for more details.

At www.packt.com, you can also read a collection of free technical articles, sign up for a range of free newsletters, and receive exclusive discounts and offers on Packt books and eBooks.

Contributors

About the author

Harish Gulati is a consultant, analyst, modeler, and trainer based in London. He has 16 years of financial, consulting, and project management experience across leading banks, management consultancies, and media hubs. He enjoys demystifying his complex line of work in his spare time. This has led him to be an author and orator at analytical forums. His published books include SAS for Finance by Packt and Role of a Data Analyst, published by the British Chartered Institute of IT (BCS). He has an MBA in brand communications and a degree in psychology.

About the reviewer

Harshil Gandhi is part of SAS India's Consulting team. As a consultant, he provides consulting and implementation services, including requirements gathering, analysis, solution development/implementation, and knowledge transfer. He also assists sales teams with relevant activities and ensures the highest levels of customer satisfaction. Harshil is good at rapidly prototyping solutions thanks to his background in data science (MTech). He is also a visiting scholar at NMIMS Mumbai.

Packt is searching for authors like you

If you're interested in becoming an author for Packt, please visit `authors.packtpub.com` and apply today. We have worked with thousands of developers and tech professionals, just like you, to help them share their insight with the global tech community. You can make a general application, apply for a specific hot topic that we are recruiting an author for, or submit your own idea.

Table of Contents

Section 2: Merging, Optimizing, and Descriptive Statistics

Preface

SAS is one of the leading enterprise tools in the world today in the fields of data management and analysis. It enables faster, easier processing of data and empowers you to get valuable business insights for effective decision-making.

This book will serve as an all-encompassing, comprehensive guide that you can refer to while preparing for your SAS certification exam. After a quick walk-through of the SAS architecture and components, this book teaches you the different ways to import and read data from different sources using SAS. You will become familiar with SAS Base, the 4GL programming language, and SQL procedures, with comprehensive coverage of topics such as data management and data analysis. You will then move on to learn about the advanced aspects of macro-programming.

By the end of this book, you will be an expert in SAS programming and will be able to handle and manage your data-related problems in SAS with ease.

Who this book is for

If you are a data professional who's new to SAS programming and wants to be an expert at it, this is the book for you. Those looking to prepare for the SAS certification exam will also find this book to be a very handy resource. Some understanding of basic data management concepts will help you get the most out of this book.

What this book covers

Chapter 1, *Introduction to SAS Programming*, introduces programming concepts and instills in you the confidence to write basic SAS programs. We will explore what happens behind the scenes in SAS and thereby ensure that the fundamentals are in place to learn advanced concepts in the book.

Chapter 2, *Data Manipulation and Transformation*, includes comprehensive coverage of data manipulation, including tasks such as numeric-to-character conversion, handling missing values and blanks, and logic and control functions.

Chapter 3, *Combining, Indexing, Encryption, and Compression Techniques Simplified*, will focus on understanding the pros and cons of various data table combination techniques. We will explore the pros and cons of techniques using examples and look under the hood to see how SAS processes code.

Chapter 4, *Power of Statistics, Reporting, Transforming Procedures, and Functions*, looks at built-in SAS procedures that help reduce the coding effort required on your part and provide you with the ability to transform data, produce statistics, run statistical tests, and produce reports.

Chapter 5, *Advanced Programming Techniques – SAS Macros*, focuses on understanding the concept of loops and SAS macros. This chapter will help you move to advanced programming within SAS.

Chapter 6, *Powerful Functions, Options, and Automatic Variables Simplified*, focuses on mastering SAS macros using system options and functions that help debug and optimize code.

Chapter 7, *Advanced Programming Techniques Using PROC SQL*, is all about the PROC SQL procedure. We will start by understanding basic concepts such as Cartesian joins, then explore the pros and cons of using DATA steps over PROC SQL. Using various examples, we will perform multiple data tasks using PROC SQL.

Chapter 8, *Deep Dive into PROC SQL*, will introduce the unique benefits of combining our understanding of PROC SQL with macros.

Chapter 9, *Data Visualization*, covers data visualization, which is vital in the world of big data, where visual analysis is key to understanding the insights that are generated from data reporting and data mining. We will look at why visualization is important and how we can produce charts in SAS to help deliver the value that data has to offer effectively.

Chapter 10, *Reporting and Output Delivery System*, focuses on the packaging and production of data reports and insights in multiple formats and platforms.

To get the most out of this book

- Basic knowledge of SAS programming is what you need to get the most out of this book.

Download the example code files

You can download the example code files for this book from your account at `www.packt.com`. If you purchased this book elsewhere, you can visit `www.packtpub.com/support` and register to have the files emailed directly to you.

You can download the code files by following these steps:

1. Log in or register at `www.packt.com`.
2. Select the **Support** tab.
3. Click on **Code Downloads**.
4. Enter the name of the book in the **Search** box and follow the onscreen instructions.

Once the file is downloaded, please make sure that you unzip or extract the folder using the latest version of:

- WinRAR/7-Zip for Windows
- Zipeg/iZip/UnRarX for Mac
- 7-Zip/PeaZip for Linux

The code bundle for the book is also hosted on GitHub at `https://github.com/PacktPublishing/Hands-On-SAS-For-Data-Analysis`. In case there's an update to the code, it will be updated on the existing GitHub repository.

We also have other code bundles from our rich catalog of books and videos available at `https://github.com/PacktPublishing/`. Check them out!

Download the color images

We also provide a PDF file that has color images of the screenshots/diagrams used in this book. You can download it here: `http://www.packtpub.com/sites/default/files/downloads/9781788839822_ColorImages.pdf`.

Conventions used

There are a number of text conventions used throughout this book.

`CodeInText`: Indicates code words in text, database table names, folder names, filenames, file extensions, pathnames, dummy URLs, user input, and Twitter handles. Here is an example: "We have now specified the desired length of the `Make` variable."

A block of code is set as follows:

```
Data Cars;
Length Make $ 15. Default=4;
Input Make $ Year;
Datalines;
Porsche_Cayenne 2018
```

Bold: Indicates a new term, an important word, or words that you see onscreen. For example, words in menus or dialog boxes appear in the text like this. Here is an example: "The process of breaking this information is called **tokenization**."

Warnings or important notes appear like this.

Tips and tricks appear like this.

Get in touch

Feedback from our readers is always welcome.

General feedback: If you have questions about any aspect of this book, mention the book title in the subject of your message and email us at customercare@packtpub.com.

Errata: Although we have taken every care to ensure the accuracy of our content, mistakes do happen. If you have found a mistake in this book, we would be grateful if you would report this to us. Please visit www.packtpub.com/support/errata, selecting your book, clicking on the Errata Submission Form link, and entering the details.

Piracy: If you come across any illegal copies of our works in any form on the Internet, we would be grateful if you would provide us with the location address or website name. Please contact us at copyright@packt.com with a link to the material.

If you are interested in becoming an author: If there is a topic that you have expertise in and you are interested in either writing or contributing to a book, please visit authors.packtpub.com.

Reviews

Please leave a review. Once you have read and used this book, why not leave a review on the site that you purchased it from? Potential readers can then see and use your unbiased opinion to make purchase decisions, we at Packt can understand what you think about our products, and our authors can see your feedback on their book. Thank you!

For more information about Packt, please visit packt.com.

Section 1: SAS Basics

This part introduces the reader to the SAS environment—writing your first program and providing a glimpse into how SAS works in the background to execute the program. Readers will learn how to manipulate and transform data using a variety of functions.

This section comprises the following chapters:

- Chapter 1, *Introduction to SAS Programming*
- Chapter 2, *Data Manipulation and Transformation*

Introduction to SAS Programming

In this chapter, we will learn and master basic SAS programming techniques. For the uninitiated of you, this chapter should be a stepping stone to SAS programming. For experienced SAS programmers, this chapter will help you revise some behind the scenes functionalities and tricks of SAS. In either case, this chapter will lay the foundation for how good an advanced SAS programmer you can be. As we progress through this chapter, we will cover the following topics:

- SAS dataset fundamentals
- SAS programming language—basic syntax
- SAS LOG
- Dataset options
- SAS operators
- Formats
- Subsetting datasets

SAS dataset fundamentals

The SAS dataset contains values that are organized as rows and columns that can be processed (read/written) by SAS. The dataset can be a data file (table) or view. Either way, a dataset is typically rectangular in format. The dataset has a descriptor portion and data portion. While in the following table, we can only see the column/variable names, the descriptor portion holds further information such as the number of rows (more commonly referred to as observations) in the dataset, date and time of creation, and the operating environment in which it was created. This section is called the **data portion**, which holds all the data values:

Cost of Living in Major Cities									
Obs	City	Index	Prev_yr_index	Housing	Food	Travel	Utility	Education	Leisure	Other
1	Adelaide	85	83	35	10	10	9	14	10	12
2	Beijing	90	92	40	10	15	10	18	5	2
3	Copenhagen	65	64	25	15	10	10	12	12	16
4	Doha	56	50	30	15	5	10	10	20	10
5	Dubai	75	76	30	16	14	10	20	8	2
6	Dublin	45	43	30	10	8	12	10	15	15
7	Hong Kong	83	88	45	5	10	15	15	9	1
8	Johannesburg	35	40	45	5	5	15	15	10	5
9	Manila	41	42	25	10	15	15	20	10	5
10	Moscow	48	53	40	20	5	5	10	10	10
11	Mumbai	83	85	40	10	15	15	10	9	1
12	Munich	65	64	35	10	10	10	10	10	15
13	New York	89	85	40	10	15	10	20	5	5
14	Oslo	60	58	25	15	5	5	15	20	16
15	Paris	70	70	30	10	5	10	10	20	15

The maximum number of observations that can be counted for a SAS dataset is determined by the long integer data type size for the operating environment. In operating environments with a 32-bit long integer, the maximum number is $2^{31}-1$ or approximately 2 billion observations (2,147,483,647). In operating environments with a 64-bit long integer, the maximum number is 263-1 or approximately 9.2 quintillion observations. Operating machines with a 32-bit long integer are likely to reach the maximum observation count of 2 billion observations in some real-world scenarios. However, the 64-big long integer machines are unlikely to ever reach the upper limit of observations permitted.

While dealing with SAS data, we are less concerned with the exact number of observations. It doesn't matter whether they are 5 million or 6 million observations. However, it should be much faster to query a 500-observation table compared to one with 5 million observations. The observations merely help in estimating the processing time. Throughout the book, we will learn about programming techniques that will help speed up processing. In this chapter, we will learn about compression.

The aspect more important than observations is the number of records per ID variable. In the cost of living table, we have 15 observations. Each observation is a record of a different city. In this case, the variable city has become an ID variable. In a transactional table of retail sales, you may have hundreds of records for each loyalty card. The multiple records may represent the basket of goods that have been purchased over a period of time. All the records would be linked to a single loyalty card number.

Data is seldom fully populated for each variable. For example, a data table constructed using responses from a questionnaire may have missing responses from a few respondents if the question that's being asked isn't mandatory. This information may not be available for each variable and such instances would be set to missing values in the table. A period (.) represents a missing numeric record, whereas a space (" ") represents a missing character record.

 Please remember that a 0 value and missing values aren't the same.

Creating an SAS table

The task that SAS performs to create a table can be categorized into two phases:

- Compile
- Execute

The following flowchart shows us the detailed compilation and execution process in table creation:

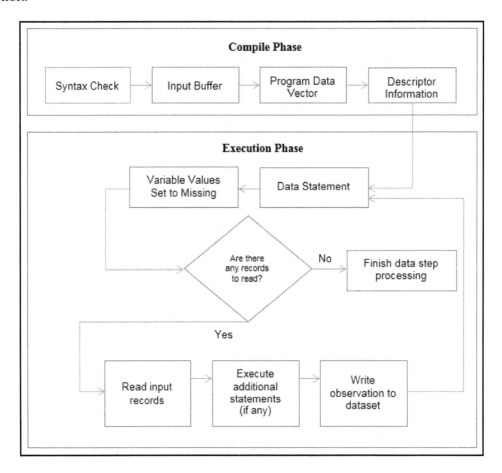

Compile phase

The compile phase is one that is often not well understood by the users as this is the backend processing the output that was generated. Tasks within the compile phase include syntax check, input buffer (not created if reading an existing dataset), **program data vector** (**PDV**), and descriptor information:

- **Syntax check**: In the syntax check task, SAS checks whether the code syntax is correct and then converts the programming statements into machine code to help execute the code. Only if the syntax is correct does SAS proceed with other tasks in the compile phase.

- **Input buffer**: The input buffer is a logical area in memory in which SAS reads each record of data from a raw data file where the program executes. In the case when a dataset is created from another SAS dataset, an input buffer is not created.
- **PDV**: This is a logical area of memory where SAS builds the dataset by writing each observation one at a time. Data is read from the input buffer. Values are assigned to the variables in the PDV. The values are written to the dataset as a single observation. There are two automatic variables created in PDV, namely _N_ and _ERROR_. Both these variables are not part of the output dataset that's created. The _N_ variable signifies the number of iterations of the data step. The _ERROR_ variable captures the number of instances in each data step when an error occurs.
- **Descriptor information**: This contains information about the dataset and includes both the dataset and variable attributes.

Execution phase

In this phase, SAS writes the PDV values to the output dataset for the current observation. The values of the PDV are set to missing. If there are any more records to read, then the program goes back to the top of the data step and executes it again. The next observation is built and stored in the output dataset. This process goes on until there are no more records to read. After this, the dataset is then closed and SAS goes to the next DATA or PROC step (if available) in the program file.

Dataset creation example

Let's look at the steps in the compile and execution phase while creating the cost of living dataset we showcased in the preceding screenshot. We will run the following program to create the dataset:

```
DATA COST_LIVING;
INPUT City $12. Index Prev_yr_index Housing Food Travel Utility Education
Leisure Other;
DATALINES;
Adelaide 85 83 35 10 10 9 14 10 12
Beijing 90 92 40 10 15 10 18 5 2
Copenhagen 65 64 25 15 10 10 12 12 16
Doha 56 50 30 15 5 10 10 20 10
Dubai 75 76 30 16 14 10 20 8 2
Dublin 45 43 30 10 8 12 10 15 15
Hong Kong 83 88 45 5 10 15 15 9 1
```

```
Johannesburg 35 40 45 5 5 15 15 10 5
Manila 41 42 25 10 15 15 20 10 5
Moscow 48 53 40 20 5 5 10 10 10
Mumbai 83 85 40 10 15 15 10 9 1
Munich 65 64 35 10 10 10 10 10 15
New York 89 85 40 10 15 10 20 5 5
Oslo 60 58 25 15 5 5 15 20 15
Paris 70 70 30 10 5 10 10 20 15
Seoul 73 75 30 10 10 10 15 15 10
Singapore 75 74 35 15 10 10 20 5 5
Tokyo 87 85 40 15 10 5 15 14 1
Zurich 63 61 30 10 10 15 10 10 15
;
RUN;
```

In its current form, the program will execute without errors. This is because the first phase of compile that checks for syntax errors will not come across any coding errors. For illustration purposes, we can try and remove the ; after the DATALINES command. The following error will be encountered when we try to run the modified code and no output table will be generated:

```
1          OPTIONS NONOTES NOSTIMER NOSOURCE NOSYNTAXCHECK;
72
73         DATA COST_LIVING;
74         Input City $12. Index  Prev_yr_index Housing Food Travel Utility Education Leisure
75         Datalines
76         Adelaide    85 83 35 10 10 9 14 10 12
           _____
           22
           76
ERROR 22-322: Syntax error, expecting one of the following: ;, CANCEL, PGM.

ERROR 76-322: Syntax error, statement will be ignored.
```

Let's review the steps in SAS processing for the preceding date creation program to understand how the PDV is generated. After the syntax check is done, the input buffer and the PDV are created. The PDV contains all the variables that are declared in the input statement. Initially, all the variable values are set to missing. The automatic _N_ and _ERROR_ variables are both set to 0:

```
Input Buffer
----+----1----+----2----+----3----+----4----+----5----+----6----+----7----
+----8----+----9----+-
```

PDV:

City	Index	Prev_yr_index	Housing	Food	Travel	Utility	Education	Leisure	Other
	.	.	.	.	.	.	.	.	.

The `City` variable has been declared as a character variable and the rest of the variables are numeric. The missing character values are written as *blanks* and the missing numeric values are written as *periods*. After this stage, the data step executes and the data values are first assigned to the input buffer before being written to the PDV:

```
Input Buffer
----+----1----+----2----+----3----+----4----+----5----+----6----+----7----
+----8----+----9----+-
Adelaide 85 83 35 10 10 9 14 10 12
```

PDV:

City	Index	Prev_yr_index	Housing	Food	Travel	Utility	Education	Leisure	Other
Adelaide	85	83	35	10	10	9	14	10	12

At this point, SAS writes the data values in the PDV to the output dataset. The _N_ variable is set to 1 and _ERROR_ is set to 0. The PDV is set to missing values. Since we have more lines of data to read, the program will keep executing:

```
Input Buffer
----+----1----+----2----+----3----+----4----+----5----+----6----+----7----
+----8----+----9----+-
Adelaide 85 83 35 10 10 9 14 10 12
```

PDV:

City	Index	Prev_yr_index	Housing	Food	Travel	Utility	Education	Leisure	Other
	.	.	.	.	.	.	.	.	.

For the second line of data, the following values in the input buffer and PDV will be written:

```
Input Buffer
----+----1----+----2----+----3----+----4----+----5----+----6----+----7----
+----8----+----9----+-
Beijing 90 92 40 10 15 10 18 5 2
```

PDV:

City	Index	Prev_yr_index	Housing	Food	Travel	Utility	Education	Leisure	Other
Beijing	90	92	40	10	15	10	18	5	2

The observations will now be written to the dataset and the variable _N_ will be incremented by 1 to take its value to 2. The _ERROR_ variable will again be reset to 0 as no errors have been encountered. This process will continue until the time the last data observations have been read by the program and sent to the output dataset.

SAS programming language – basic syntax

We used code in the first program to create an output dataset. The dataset was created by what is known as **data steps**. Data steps are a collection of statements that can help create, modify, and control the output. SAS also leverages **Structured Query Language (SQL)**. Let's review the basic syntax for data steps and SQL within SAS. We will continue to explore more advanced versions of code throughout this book.

Data step

The following code represents one of the simplest forms of a data step:

```
DATA WORK.Air;
SET SASHELP.Air;
RUN;
```

Using the SET statement in this data step, we have specified the dataset that we want to refer to. There are no conditional statements in the program that are selecting a proportion of records from the Air dataset in the SASHELP library. As a result, all the contents of the dataset in the set statement will get copied over to the dataset specified in the data statement. In the set statement, it is necessary to specify a dataset. However, if you specify _LAST_ in the SET statement, the dataset that was last created in the SAS session will be used instead. Finally, the Run command is specified to execute the program. The only instance when the command isn't needed in a data statement is when creating a dataset using an INPUT statement (as shown in the following code block).

The use of DATA signifies that we are using the data step. In this statement, we specify the output dataset. WORK is what is known as a **library** in SAS. A library is like a Windows folder that stores files and various other things such as formats and catalogs. Every SAS session (each instance of a SAS software invocation is a separate session) is assigned its own temporary workspace. The temporary workspace is known as the **Work library**. At the end of the session, the temporary work session is cleared and unless saved in a permanent library, all the contents of the Work library are deleted. If the Work library is not specified, the dataset will be created in the temporary workspace by default.

A permanent library is one that can be assigned using a physical path. In BASE SAS, this can be a physical folder located in the computer drives. In the case of SAS Enterprise Guide Studio and other software, this may be a space on the server. The dataset name consists of two parts—the library name followed by the dataset name. Both are separated by a period. For the creation of datasets in the Work library, users only need to specify the dataset name. If no dataset name is specified, SAS names the dataset as D1, D2, and so on:

```
DATA;
INPUT Id;
DATALINES;
1
2
;
RUN;
```

Since this is the first program in our SAS session without a dataset name specified, the name *D1* will be assigned to the dataset:

```
1            OPTIONS NONOTES NOSTIMER NOSOURCE NOSYNTAXCHECK;
72
73           data;
74           input id;
75           datalines;

NOTE: The data set WORK.DATA1 has 2 observations and 1 variables.
NOTE: DATA statement used (Total process time):
      real time              0.00 seconds
      cpu time               0.00 seconds
```

We will explore the data step options in further detail throughout this book.

Proc SQL

SAS leverages SQL through a built-in procedure. While we aren't going to focus on SQL in this book, let's look at the basic structure of a SQL query in SAS:

```
PROC SQL;
   CREATE TABLE Table_Name AS
   SELECT
   FROM
   WHERE
   GROUP BY;
QUIT;
```

PROC is the command that's used to specify a built-in procedure in SAS. In the case of the preceding program, we are referring to the SQL procedure. Just like in the data step, we start off by specifying the table name that we are creating. We then list the variables that we want to select from another dataset. The name of the dataset that has been selected is named in the FROM statement. The WHERE clause is used to sub-select the data. The GROUP BY clause is used for summary functions. Finally, we end the procedure by specifying the QUIT argument.

SAS LOG

The SAS LOG section of your coding environment is where all the actions performed that have been by the user in the current session are stored. These include instances of program submission and also messages about any programs that you might have terminated while they were executing. Apart from these, the SAS LOG also contains system-generated messages. These are of two types. The first instance is where the SAS version and a few other details about your system are written to the LOG. The second is when responses to the user code are generated and written to the LOG. The response could state that the program has run successfully, failed, or has some syntax issues that have been ignored. The responses are categorized in NOTE, INFO, WARNING, and ERROR categories. Program submission messages can be easily identified in the LOG as they have a line number associated with them.

Let's examine the LOG that's generated after running the first program from the preceding *Data step* section:

```
1              OPTIONS NONOTES NOSTIMER NOSOURCE NOSYNTAXCHECK;
72
73             DATA WORK.Air;
74             SET SASHELP.Air;
75             RUN;

NOTE: There were 144 observations read from the data set SASHELP.AIR.
NOTE: The data set WORK.AIR has 144 observations and 2 variables.
NOTE: DATA statement used (Total process time):
      real time           0.00 seconds
      cpu time            0.00 seconds


76
77
78             OPTIONS NONOTES NOSTIMER NOSOURCE NOSYNTAXCHECK;
91
```

On the left-hand side of the LOG is the line number. Line number 1 contains the default settings that are in place for this SAS session. The log for our program starts getting generated in line 73. From line 73 to line 75, the program that has been specified in the program editor window is replicated in the log. There are no errors being produced in the log, unlike the one shown in the *Data creation example* section. The note that's produced mentions the number of observations read from the input dataset. It also contains the number of observations and variables in the output dataset. The time it took to execute the query is also mentioned at the end of the notes.

By reading the LOG, the user can review the program executed and notes, warnings, or errors produced, review the summary produced about the input and output dataset, and check query execution time. After the first run, the user may want to modify the program. This could be because the output is not in sync with the intended requirement or an error has been generated. In any case, understanding of the log is required before we can edit the program.

Naming conventions in SAS

Some of the frequently used functionalities in SAS where naming conventions need to be followed are variables, datasets, formats or informats that are user-created, arrays, labels, macro variables, library names, and file references.

The general rules for SAS names are as follows:

- Variable names can be up to 32 characters, whereas some other names such as library names can be up to 8 characters.
- SAS isn't case sensitive in name specification, unlike some other programming languages or other statistical packages.
- The name cannot start with a number. It needs to start with a letter or an underscore. The second character can be a number or an underscore. No special characters apart from underscore are allowed in the name. Underscores are frequently used by programmers for variable names where multiple words are involved, for example, `Order_Date`, `Payment_Date`, or `Delivery_Date`. A variable name cannot contain a blank. Hence, the underscore becomes an important way to make the variable names more legible for users of your code and data.
- In some instances of SAS names for `filerefs`, some special characters are allowed.
- Some names are reserved for SAS functions and keywords that are used by the system. For instance, you cannot specify a library name that is the default SAS library associated with your SAS installation. These include SASHELP, SASUSER, and WORK.

SAS already has a macro variable called `sysdate`. Users shouldn't attempt to create a macro variable with the same name.

 The maximum length of arrays, labels, variables, numeric formats, macros, and datasets is 32. Character formats have a length of 31. Character and numeric informats have a length of 30 and 31, respectively. File references and library names have a maximum length of 8.

Naming conventions for Teradata in SAS

The Teradata naming conventions are different from the SAS names. Some of the key aspects are the following:

- Unlike SAS, where the name can be up to 32 characters, Teradata names in SAS need to be between 1 and 30 characters.
- The name can contain the letters A to Z, numbers, underscores, and also the dollar and pound signs.

- You can specify a name in double quotes. That way, it can contain any characters except double quotation marks.
- Names in double quotes are not case sensitive.

Dataset options

There are many built-in SAS options that apply to the dataset. The broad purpose of these options is to help us do the following:

- Rename variables
- Select variables for subsetting
- Retain select variables
- Specify the password for a dataset, compress it, and encrypt it

Throughout this book, we will be looking at various dataset options. We will begin by exploring the compress, encrypt, and index options.

Compression

Compression reduces the number of bytes that are required to store each observation. The advantage of compression is that it requires less storage, owing to the fact that fewer bytes are required and fewer I/O operations are necessary to read or write to the data during processing.

The biggest disadvantage of compression is that more CPU resources are required to read a compressed file and there are situations where the resulting file size might increase rather than decrease the time that's required to execute SAS statements. This is one of the reasons why many SAS users end up not compressing datasets and miss out on the advantages. Users need to be aware that some SAS administrators are known to make compression the default option on a server level.

The COMPRESS=YES dataset helps compress a file. This option works with a SAS data file and not a view. This effects only the dataset specified in the output statement and not the dataset in the SET statement. The best way to uncompress a file is to create a new file. Alternatively, you can use COMPRESS=NO while recreating the dataset. Remember that the YES option remains activated for the rest of your SAS session unless the NO option is specified.

The benefit of compression can be gauged by looking at the SAS LOG after compression. A message is generated in the LOG stating the percentage reduction in size of the compressed file compared to the size that would have been in an uncompressed state:

```
DATA WORK.Air (COMPRESSION = YES);
SET SASHELP.Air;
RUN;
```

The following note is written to the SAS log, specifying that compression was disabled as it isn't advantageous:

```
NOTE: Compression was disabled for data set WORK.AIR because compression
overhead would increase the size of the data set.
```

Here, we know the dataset was produced without utilizing the compress option.

Encryption

While encrypting a dataset, you have to use the READ or the PW dataset option. If the password is lost, then only the SAS helpdesk can resolve the situation. The only other way to change the password is to recreate the dataset. We will attempt to encrypt the Air dataset using the following code block:

```
DATA WORK.Air (ENCRYPT = YES READ=CHAPTER2);
SET SASHELP.Air;
RUN;

PROC PRINT DATA = WORK.Air(READ=CHAPTER2);
RUN;
```

We will have to specify the password to be able to read the dataset using the print procedure.

Indexing

As the name suggests, indexing is a way to tell SAS how things have been arranged. Typically, the SAS dataset is stored in pages. There is no way for SAS to know on which page or sequence the information is stored unless the dataset is indexed. The index stores values in ascending value order for a specific variable or variables and includes information as to the location of those values within observations in the data file. In other words, an index allows you to locate an observation by value.

Indexes can be done on one variable or multiple variables. If done on multiple variables, they are called **composite indexes**. Indexes needed to be stored somewhere for the subsequent programs using the indexed data to be able to leverage them. They are stored as a separated data file and all indexes of a dataset are stored in a single file. Indexes aren't automatically transferred to the dataset that is created using an indexed input dataset. However, if you add or delete observations in the indexed dataset, the index is automatically updated. The good thing about indexes is that they can be created at the stage of creating a dataset or at any time afterwards. The dataset doesn't need to be compressed for indexing.

The pros of creating an index are as follows:

- For WHERE processing, an index can provide faster and more efficient access to a subset of data.

 Note that to process a WHERE expression, SAS by default decides whether to use an index or to read the data file sequentially. Please note that the IF condition never leverages the index.

- For BY processing, an index returns observations in the index order, which is in ascending value order, without using the SORT procedure, even when the data file is not stored in that order.

The biggest disadvantage of index creation is the cost associated with it. Storage of the runtime and the time to create the index are big overheads. By default, indexes should never be created. These should only be created if the dataset in question is going to be queried repeatedly. Another clear instance when an index should be created is if its presence is significantly reducing the runtime associated with the analytical tasks the users are performing. The user should be prudent about which variables in the dataset need an index.

The syntax for index creation is the following:

```
INDEX CREATE index-specification-1 <...index-specification-n>
  </ <NOMISS> <UNIQUE> <UPDATECENTILES= ALWAYS | NEVER | integer>>;
```

Arguments for the syntax are index-specification(s).

It can be one or both of the following forms:

- `variable`: Creates a simple index on the specified variable
- `index=(variables)`: Creates a composite index

The optional arguments are the following:

- `NOMISS`: Excludes all observations with missing values for all index variables from the index
- `UNIQUE`: Specifies that the combination of values of the index variables must be unique
- `UPDATECENTILES=ALWAYS | NEVER | integer`: Specifies when centiles are to be updated

Let's create a simple index using the following code block:

```
PROC DATASETS LIBRARY=WORK;
    MODIFY Cost_Living;
        INDEX CREATE City;
RUN;
```

On running, the following messages are written to the log:

```
NOTE: Simple index City has been defined.
NOTE: MODIFY was successful for WORK.COST_LIVING.DATA.
```

To create a composite index on the `City` and `Index` variables, use the following statement in `PROC DATASETS`:

```
INDEX CREATE City_and_Index = (City Index);
```

The key difference in syntax between single and composite indexes is that you have to specify a unique name for the composite index.

SAS operators

An SAS operator is a symbol that represents a comparison, arithmetic calculation, or logical operation. We will look into the various operator in the following sections.

Arithmetic operators

For performing arithmetic operations, the following can be used in SAS:

Symbol	Definition
+	Addition
/	Division
**	Exponentiation
*	Multiplication
-	Subtraction

Comparison operators

Comparison operators set up a comparison, operation, or calculation with two variables, constants, or expressions. If the comparison is true, the result is 1. If the comparison is false, the result is 0. The following table show the symbol, mnemonic equivalent, and definition of each comparison operators in SAS:

Symbol	Mnemonic equivalent	Definition
=	EQ	Equal to
^= (or ¬=, ~=)	NE	Not equal to
>	GT	Greater than
<	LT	Less than
>=	GE	Greater than or equal
<=	LE	Less than or equal
	IN	Equal to one of a list

Logical operators

Logical operators, also called **Boolean operators**, are usually used in expressions to link sequences of comparisons. The logical operators are shown in the following table:

Symbol	Mnemonic equivalent
&	AND
	(or !, ¦)	OR
¬ (or ^, ~)	NOT

Formats

There are usually three scenarios when formats are used in SAS:

- To format the data available to make it more readable
- To assign a specific format to make the data more meaningful
- To alter the data type to make calculations, derivations, and so on

Formatting to make the data readable

SAS stores all dates as single unique numbers in a numeric format. All dates are stored as the number of dates from January 1, 1960. If you think of a number line, all dates prior to January 1, 1960 are negative and all dates after that are positive. It's cumbersome to decode the numeric dates. An easier alternative is to get SAS to display a date using a date format. Let's look at our cost of living dataset with the addition of a new date variable:

```
DATA COST_LIVING;
INPUT City $12. Index Prev_yr_index Housing Food Travel Utility Education
Leisure Other Updated MMDDYY6.;
DATALINES;
Adelaide 85 83 35 10 10 9 14 10 12 010118
Beijing 90 92 40 10 15 10 18 5 2 010118
Copenhagen 65 64 25 15 10 10 12 12 16 020118
Doha 56 50 30 15 5 10 10 20 10 030118
Dubai 75 76 30 16 14 10 20 8 2 040118
.
.
.
;
```

The input data is available in a month, date, and year format represented by six numbers. To ensure that SAS can store this date as a number, we need to specify what is known as an **informat**. Here, MMDDYY6. is an informat with the length of the variable as six digits:

Obs	City	Index	Prev_yr_index	Housing	Food	Travel	Utility	Education	Leisure	Other	Updated
1	Adelaide	85	83	35	10	10	9	14	10	12	21185
2	Beijing	90	92	40	10	15	10	18	5	2	21185
3	Copenhagen	65	64	25	15	10	10	12	12	16	21216
4	Doha	56	50	30	15	5	10	10	20	10	21244
5	Dubai	75	76	30	16	14	10	20	8	2	21275

In the preceding table, we can see that the date of index value in `Updated` for `Adelaide` is `21185`. But is this the value that we get when we do January 1, 2018 – January 1, 1960?

A leap year occurs every 4 years, except for years that are divisible by 100 and not divisible by 400.

So, the mean length of the Gregorian calendar year is as follows:

1 mean year = (365+1/4-1/100+1/400) days = 365.2425 days

There are 58 years between the update for Adelaide and the start of the SAS date value. This gives us a number of *58 x 365.2425 = 21184.065*. This is close to the value of 21185 we got. In fact, if we multiply the number of years (58) by the sidereal year (the time it takes for Earth to do a single rotation around the Sun) value of 365.25636, we get a value of 21184.86888. This value, when rounded up, gives us the numeric value of 21185 for the updated variable for the first observation.

If you had the date value of January 1, 1960, SAS will store it as 0. January 2, 1960, will be stored as 1. However, December 31, 1959 will be stored as -1. Please don't get confused when you see negative date values. All you need to do is specify a date format to make the value readable.

We will now add the following `Format` statement between the `INPUT` and `DATALINES` arguments to the preceding code:

```
Format Updated Date9.;
```

This produces an easily readable date value. Remember to add a date format when dealing with dates. An input format is also required to ensure SAS correctly interprets the date:

Obs	City	Index	Prev_yr_index	Housing	Food	Travel	Utility	Education	Leisure	Other	Updated
1	Adelaide	85	83	35	10	10	9	14	10	12	01JAN2018
2	Beijing	90	92	40	10	15	10	18	5	2	01JAN2018
3	Copenhagen	65	64	25	15	10	10	12	12	16	01FEB2018
4	Doha	56	50	30	15	5	10	10	20	10	01MAR2018
5	Dubai	75	76	30	16	14	10	20	8	2	01APR2018

Specifying a format to make it meaningful

Consider the example of a **management information** (**MI**) reporting analyst based in California. The analyst will have to extract information about the sales performance of each region from the central data warehouse to present a global sales performance MI report.

Should the analyst be expected to intuitively know in which currency the revenue of each region is stored in the warehouse? It would be helpful if formats were assigned to the revenue to signify whether the local or a commonly used currency (such as the US Dollar, Yen, Euro, or British Pound) has been used to store the regional revenue figures.

The following statement, when added between the `DATA` and `SET` statements, converts the numerical index value into dollars:

```
FORMAT Index dollar6.2;
```

In the preceding snippet, the `.2` represents the two decimal spaces that have been specified. `6` is the length of the format, which includes the dollar symbol. If we had specified 5.2 as the length of the format, then the output value would have been in the format 85.00 instead of $85.00. The syntax of the format is *w.d*, where *w* is the width of the format and *d* is the number of decimal places required. Specifying the format allows for adequate space to write the number along with the decimal specification, a period, and a negative sign. The width of the format can take up any number from 1-32. The decimal specification should be less than the width. If you don't specify a decimal specification, then the value will be written without the decimal point by default.

The following is the output that's produced after specifying the dollar format:

Obs	Index	City	Prev_yr_index	Housing	Food	Travel	Utility	Education	Leisure	Other	Updated
1	$85.00	Adelaide	83	35	10	10	9	14	10	12	01JAN2018
2	$90.00	Beijing	92	40	10	15	10	18	5	2	01JAN2018
3	$65.00	Copenhagen	64	25	15	10	10	12	12	16	01FEB2018
4	$56.00	Doha	50	30	15	5	10	10	20	10	01MAR2018
5	$75.00	Dubai	76	30	16	14	10	20	8	2	01APR2018

The syntax for some of the other currency formats is as follows:

```
FORMAT Index Euro7.2;/*Euro*/
FORMAT Index NLMNLGBP9.2; /*British Pound*/
FORMAT Index NLMNLJPY9.2; /*Japanese Yen*/
FORMAT Index NLMNLAUD9.2; /*Australian Dollar*/
FORMAT Index NLMNLNZD9.2; /*New Zealand Dollar*/
/*This is a comment*/;
```

In SAS, comments are written between '/* */' or, at the start of a statement, a comment can be specified by prefixing an asterisk before a statement. If prefixing an asterisk, remember to end the sentence with a semicolon, otherwise else the next statement will also be commented by default and will not execute.

Altering the data type

While processing data, you will frequently need to change data types. Even if the data is of good quality, where numbers have been stored as numerical and all date fields have been specified with the correct date format, transformations or derivations from them may be required to perform analysis.

Some of the commonly used transformations and derivations for a data analyst where the data type is altered are the following:

- Converting from numerical into character
- Converting from character into numerical
- Using a date or datetime format
- Extracting information from data

To convert a number into a character, use the PUT function. The syntax is as follows:

```
Character_variable = put(numeric_variable, informat.);
```

To convert the index variable in the cost of living dataset into a character, use the following code:

```
DATA Num_to_Char;
SET Cost_Living;
Index_char = PUT(Index, 3.);
RUN;
```

First, we converted a variable from numeric into character format and then added a new variable to the dataset.

To convert from character to numeric, we use the INPUT function. The syntax is as follows:

```
Numeric_variable = input(character_variable, informat.);
```

Let's run the following code to create a dataset that will help us understand the character into numeric conversion:

```
DATA Convert;
INPUT Id_Char $4. Turnover $7. Turnover_w_Currency $8. Source_Mixed $3.;
DATALINES;
0001 20,000 $20,000 A1
0002 10,000 $10,000 2
;
```

This gives us the following dataset:

Obs	Id_Char	Turnover	Turnover_w_Currency	Source_Mixed
1	0001	20,000	$20,000	A1
2	0002	10,000	$10,000	2

To convert the `Id_char` variable to numeric, we can use the following statement in a data step:

```
Id_Num = INPUT(Id_Char, 5.);
```

This will remove the leading zeros and give us the values 1 and 2 for the variable. But what if we wanted to preserve the leading zeros? In that case, we should use the following code:

```
Id_Num_Leading_Zero = INPUT(Id_Char, $5.);
```

To convert the `Turnover_w_currency` variable to numeric data type, we need to use the correct informat:

```
Turnover_Num = INPUT(Turnover_w_currency, dollar8.);
```

The `Source_mixed` field has alphanumeric data. When we use the following statement to convert it into a numeric variable, we get missing data for observation 1. Users should be careful while deploying numeric conversion using automated scripts. Missing data can be deemed acceptable in some circumstances. However, losing the alphanumeric data might negatively impact data quality in some instances:

```
Source_numeric = INPUT(source_mixed, 3.);
```

We have looked at date values and how SAS interprets them as numbers. The date variable with time information is also stored as a number in SAS. The informat and format used for interpretation are different from those used for a date variable. We will try and understand handling date-time values using the following example:

```
DATA DateTime;
INPUT Id Date_Time Datetime20.;
DATALINES;
1 01aug19:09:10:05.2
2 01aug20:19:20:10.4
;

DATA Convert_DateTime;
SET DateTime;
FORMAT Orig_Date Datetime.;
Orig_Date = Date_Time;
```

```
FORMAT Orig_Date_1 Datetime7.;
Orig_Date_1 = Date_Time;
FORMAT Orig_Date_2 Datetime12.;
Orig_Date_2 = Date_Time;
RUN;
```

The datetime values must be in the following form: ddmmmyy or ddmmmyyyy, followed by a blank or special character, followed by hh:mm:ss.ss (the time), where hh is the number of hours ranging from 00 through 23, mm is the number of minutes ranging from 00 through 59, and ss.ss is the number of seconds ranging from 00 through 59 with the fraction of a second following the decimal point. Only the ss.ss portion is optional in the datetime value.

We get the following output when we create the DateTime table using informats:

Obs	Id	Date_Time
1	1	1880269805.2
2	2	1911928810.4

Without the time variable, the values for IDs **1** and **2** (dates August 01, 2019 and August 01, 2020) will be stored in SAS as 21762 and 22128. The addition of the time component to the date changes the value of the number stored in SAS significantly.

To make the datetime information stored in SAS readable, we used different formats in DATA Convert_DateTime and got the following output:

Obs	Id	Date_Time	Orig_Date	Orig_Date_1	Orig_Date_2
1	1	1880269805.2	01AUG19:09:10:05	01AUG19	01AUG19:09
2	2	1911928810.4	01AUG20:19:20:10	01AUG20	01AUG20:19

For date and time values, we have looked at how to read and output them in SAS. We have also managed to format these dates in a way that we only see them in a manner relevant to us. On occasions, we might just want to extract some information from the date or the datetime variable. This information could be stored in a non-date/datetime variable such as the numeric variable.

We can use the following statements to extract the `Year`, `Month`, and `Date` values from a `Date` variable with an informat of `mmddyy`:

```
Year = YEAR (Date);
Month = MONTH (Date);
Day = DAY (Date);
```

For the date August 1, 2019, we will get the year as 2019, month as 8, and day as 1 when we use the preceding statements in a data step. The `Year`, `Month`, and `Day` variables will be created as numeric variables.

If we have a `Date_Time` variable, we need to extract the date part from it before converting it into a `Year`, `Month`, and `Date` variable. We can use the following statements:

```
Year = YEAR (DATEPART(Date_Time));
Month = MONTH (DATEPART(Date_Time));
Day = DAY (DATEPART(Date_Time));
```

For the datetime value `01aug19:09:10:05.2`, we will get the values `2019`, `8`, and `1` for the year, month, and day variables respectively.

Subsetting datasets

In many instances, we will need to select only a proportion of the overall dataset. This selection or subsetting of the dataset can be done in three ways:

- Use of the `WHERE` and `IF` statements
- Using the SAS dataset `OPTIONS`
- Using the `DROP` and `KEEP` statements

WHERE and IF statements

Let's only select the data where the records have been updated in `2019`:

```
DATA Updated_2019;
SET Cost_Living;
WHERE Year(Updated) = 2019;
RUN;
```

The following output will be generated:

Obs	City	Index	Prev_yr_index	Housing	Food	Travel	Utility	Education	Leisure	Other	Updated
1	Manila	41	42	25	10	15	15	20	10	5	01JAN2019
2	Moscow	48	53	40	20	5	5	10	10	10	01JAN2019
3	Mumbai	83	85	40	10	15	15	10	9	1	01JAN2019
4	Munich	65	64	35	10	10	10	10	10	15	01JAN2019
5	New York	89	85	40	10	15	10	20	5	5	01JAN2019
6	Oslo	60	58	25	15	5	5	15	20	15	01JAN2019
7	Paris	70	70	30	10	5	10	10	20	15	01JAN2019
8	Seoul	73	75	30	10	10	10	15	15	10	01JAN2019
9	Singapore	75	74	35	15	10	10	20	5	5	01JAN2019
10	Tokyo	87	85	40	15	10	5	15	14	1	01JAN2019
11	Zurich	63	61	30	10	10	15	10	10	15	01JAN2019

In this instance, we could have also used the IF statement to generate similar output. We can also use WHERE and IF statements in the same data step:

```
DATA Updated_2019;
SET Cost_Living;
WHERE Year(Updated) = 2019;
IF Index >= 80;
RUN;
```

This further subsets the data and produces the following output:

Obs	City	Index	Prev_yr_index	Housing	Food	Travel	Utility	Education	Leisure	Other	Updated
1	Mumbai	83	85	40	10	15	15	10	9	1	01JAN2019
2	New York	89	85	40	10	15	10	20	5	5	01JAN2019
3	Tokyo	87	85	40	15	10	5	15	14	1	01JAN2019

While subsetting, we should be aware of whether the WHERE or IF statement will be applicable in a data step. The WHERE statement executes at the PDV level, that is, only the selected/subset data will be read in the PDV. This can considerably reduce the time to execute a query. The IF statement can only be executed after all the data has been read into the PDV. The WHERE statement should be preferred over the IF statement if the dataset being read is indexed. The WHERE statement allows SAS to directly retrieve the specified value from the indexed table.

Because the WHERE statement executes at the PDV level, it cannot leverage any new variables that are created in the data step. The following code block is an example of the incorrect use of the WHERE statement:

```
DATA Known_Components;
SET Index;
Known_Component_Index = Index-Other;
WHERE Known_Component_Index >= 80;
RUN;
```

This produces the following error in the Log:

```
1          OPTIONS NONOTES NOSTIMER NOSOURCE NOSYNTAXCHECK;
72
73         Data Known_Components;
74         Set Cost_Living;
75         Known_Component_Index = Index-Other;
76         WHERE Known_Component_Index >= 80;
ERROR: Variable Known_Component_Index is not on file WORK.COST_LIVING.
77         Run;
```

We can instead replace the WHERE statement with the following IF statement:

```
IF Known_Component_Index >= 80;
```

The program executes without any errors and the output dataset is created with five observations, where known_component_index is greater than or equal to 80.

Using OPTIONS

We can restrict the number of records being read using the OBS and the FIRSTOBS options:

```
PROC PRINT DATA = Cost_Living (FIRSTOBS=4 OBS=5);
RUN;
```

This will give us the following output:

Obs	City	Index	Prev_yr_index	Housing	Food	Travel	Utility	Education	Leisure	Other	Updated
4	Doha	56	50	30	15	5	10	10	20	10	01MAR2018
5	Dubai	75	76	30	16	14	10	20	8	2	01APR2018

The option can also be specified outside the data step:

```
OPTIONS OBS = 2;
```

Using OPTIONS in SAS is effective for subsetting but only when the criterion is the number of records and when the starting position is observations is known or doesn't matter.

DROP or KEEP options

There will be countless occasions when a user will need to use the DROP or KEEP options. These options help retain the records that are needed in the output dataset. Using certain programming statements can help reduce the execution time of the program by subsetting the records from the input dataset. Let's look at the following example:

```
DATA Keep_and_Drop (DROP = Prev_yr_index);
SET Cost_Living (KEEP = City Index Prev_yr_index);
WHERE Index < Prev_yr_index;
RUN;
```

This produces the following output:

Obs	City	Index
1	Beijing	90
2	Dubai	75
3	Hong Kong	83
4	Johannesburg	35
5	Manila	41
6	Moscow	48
7	Mumbai	83
8	Seoul	73

The KEEP option helped restrict the input dataset being read to include only the **City**, **Index**, and Prev_yr_index variables. After this, the WHERE statement was executed. Finally, using the DROP statement, we removed the Prev_yr_index variable from the output dataset. Another way to use the KEEP option can be as a separate statement within a data step:

```
KEEP City Index Prev_yr_index;
```

Viewing properties

The content procedure shows the details of the datasets and prints the directory of the SAS library. The following is the basic form of the contents procedure:

```
PROC DATASETS Library=Work;
  CONTENTS DATA=Cost_Living;
RUN;
```

It produces a fairly descriptive output:

Directory	
Libref	WORK
Engine	V9
Physical Name	/tmp/SAS_workB9470000093A_10.0.2.15/SAS_work421F0000093A_10.0.2.15
Filename	/tmp/SAS_workB9470000093A_10.0.2.15/SAS_work421F0000093A_10.0.2.15
Inode Number	671604
Access Permission	rwx------
Owner Name	sasdemo
File Size	4KB
File Size (bytes)	4096

#	Name	Member Type	File Size	Last Modified
1	CONVERT	DATA	128KB	09/01/2019 07:55:54
2	CONVERT_DATETIME	DATA	128KB	09/01/2019 07:57:53
3	COST_LIVING	DATA	128KB	09/01/2019 08:08:23
4	DATETIME	DATA	128KB	09/01/2019 07:57:18
5	FORMAT	DATA	128KB	09/01/2019 07:52:33
6	KEEP_AND_DROP	DATA	128KB	09/01/2019 08:12:17
7	REGSTRY	ITEMSTOR	32KB	09/01/2019 07:41:34
8	SASGOPT	CATALOG	12KB	09/01/2019 07:41:35
9	SASMAC1	CATALOG	208KB	09/01/2019 07:41:34
10	SASMAC2	CATALOG	20KB	09/01/2019 07:41:34
11	SASMAC3	CATALOG	20KB	09/01/2019 07:41:34
12	SASMAC4	CATALOG	20KB	09/01/2019 08:13:31
13	SASMAC5	CATALOG	20KB	09/01/2019 07:41:34
14	SASMAC6	CATALOG	20KB	09/01/2019 07:41:34
15	SASMAC7	CATALOG	20KB	09/01/2019 07:41:34
16	SASMAC8	CATALOG	20KB	09/01/2019 07:41:34
17	SASMAC9	CATALOG	20KB	09/01/2019 07:41:34
18	SASMACR	CATALOG	20KB	09/01/2019 08:12:17
19	UPDATED_2019	DATA	128KB	09/01/2019 08:09:31

The DATASETS procedure produces details of only the dataset specified in the program:

The DATASETS Procedure			
Data Set Name	WORK.COST_LIVING	Observations	19
Member Type	DATA	Variables	11
Engine	V9	Indexes	0
Created	09/01/2019 09:08:23	Observation Length	96
Last Modified	09/01/2019 09:08:23	Deleted Observations	0
Protection		Compressed	NO
Data Set Type		Sorted	NO
Label			
Data Representation	SOLARIS_X86_64, LINUX_X86_64, ALPHA_TRU64, LINUX_IA64		
Encoding	utf-8 Unicode (UTF-8)		

The engine/host-dependent information can be leveraged for optimizing the ability to store a dataset. For our purposes, this information isn't relevant as we are dealing with a small dataset:

Engine/Host Dependent Information	
Data Set Page Size	65536
Number of Data Set Pages	1
First Data Page	1
Max Obs per Page	681
Obs in First Data Page	19
Number of Data Set Repairs	0
Filename	/tmp/SAS_workB9470000093A_10.0.2.15/SAS_work421F0000093A_10.0.2.15/cost_living.sas7bdat
Release Created	9.0401M6
Host Created	Linux
Inode Number	671650
Access Permission	rw-rw-r--
Owner Name	sasdemo
File Size	128KB
File Size (bytes)	131072

The list of variables provides the type of the variable, its length, and its format information:

Alphabetic List of Variables and Attributes				
#	Variable	Type	Len	Format
1	City	Char	12	
8	Education	Num	8	
5	Food	Num	8	
4	Housing	Num	8	
2	Index	Num	8	
9	Leisure	Num	8	
10	Other	Num	8	
3	Prev_yr_index	Num	8	
6	Travel	Num	8	
11	Updated	Num	8	DATE9.
7	Utility	Num	8	

Dictionary tables

Some SAS software packages require minimal coding experience to get your work done. However, once you want to do anything advanced, it gets restrictive to use just the predefined options available in the package. At that moment, most users tend to write code. Chances are that if you want to go beyond what SAS already offers as predefined procedures and functionalities, you want to write advanced programs/macros. While we will delve into macros later on in this book, there is an aspect of SAS that is being leveraged less by some users. These are the dictionaries. But what are they and how can they be useful?

Some reasons for using dictionary tables are as follows:

- They hold all the information about SAS libraries, datasets, macros, and external files that are being used in the SAS session.
- It is a read-only view, so there are no chances of it being compromised.
- You can get the most up-to-date information about your session as every time you access the table, SAS determines the current state and shares the most up-to-date information.

- The information in the dictionary tables can help in writing advanced code or can be used as a basis to select observations, join tables, and so on.

The dictionary tables can be accessed easily using PROC SQL. They can be also accessed via the data steps but that entails referring to the PROC SQL view of the table in the SASHELP library. While we aren't going to focus on PROC SQL in the book, for ease of demonstrating the power of dictionary tables, I am going to use SQL.

In case you have doubts about using dictionary tables to view information about your own SAS session, think again. SAS sessions can get complex quickly. You may have imported files, thousands of lines of codes, or tens of process nodes in a SAS Enterprise Guide sort of package. It all stacks up pretty quickly and this is where dictionary tables come in handy. If you want to build any application on the back of your SAS session, the information that's held in these tables is useful.

The first step is to understand the structure of the indexes that are created in the dictionary tables. Run the following command:

```
PROC SQL;
Describe Table Dictionary.Indexes;
```

We get the following notes in the logs:

```
NOTE: SQL table DICTIONARY.INDEXES was created like:

  create table DICTIONARY.INDEXES
    (
     libname char(8) label='Library Name',
     memname char(32) label='Member Name',
     memtype char(8) label='Member Type',
     name char(32) label='Column Name',
     idxusage char(9) label='Column Index Type',
     indxname char(32) label='Index Name',
     indxpos num label='Position of Column in
Concatenated Key',
     nomiss char(3) label='Nomiss Option',
     unique char(3) label='Unique Option',
     diagnostic char(256) label='Diagnostic Message from File Open Attempt'
    );
```

Understanding the preceding information is critical if we want to be able to utilize dictionary tables. The notes are in the following format:

- The first word on each line is the column (or variable) name, that is, the name that you need to use when writing a SAS statement that refers to the column (or variable).
- Following the column name is the specification for the type of variable and the width of the column.
- The name that follows `label=` is the column (or variable) label.

Now, we will run a dictionary table query using the given notes. The following query has been run on a new SAS session that has no imported files, programs that have been run, or indexes that have been created:

```
PROC SQL;
        Select * From Dictionary.Tables;
QUIT;
```

Let's look at the output:

Library Name	Member Name	Member Type	DBMS Member Type	Data Set Label	Data Set Type	Date Created	Date Modified	Number of Physical Observations
WORK	CONVERT	DATA			DATA	01SEP19:08:55:54	01SEP19:08:55:54	2
WORK	CONVERT_DATETIME	DATA			DATA	01SEP19:08:57:54	01SEP19:08:57:54	2
WORK	COST_LIVING	DATA			DATA	01SEP19:09:08:23	01SEP19:09:08:23	19
WORK	DATETIME	DATA			DATA	01SEP19:08:57:19	01SEP19:08:57:19	2
WORK	FORMAT	DATA			DATA	01SEP19:08:52:34	01SEP19:08:52:34	5
WORK	KEEP_AND_DROP	DATA			DATA	01SEP19:09:12:17	01SEP19:09:12:17	8
WORK	UPDATED_2019	DATA			DATA	01SEP19:09:09:32	01SEP19:09:09:32	3

The preceding table is a snapshot of the dictionary table output.

We got hundreds of rows of output when we requested the contents of dictionary tables for our SAS session. This is because there is a lot of information stored in the SAS backend about the SAS help files and settings even before any code is run in the session. Let's leverage the information we have so far: we've learned about the structure of the dictionary tables to reduce the information we want to see. We will now run the following code in the session after creating the COST_LIVING dataset:

```
PROC SQL;
            Select * From Dictionary.Tables
            Where Libname eq "WORK"
    And Memname eq "COST_LIVING";

            Select * From Dictionary.Columns
            Where Libname eq "WORK"
            And Memname eq "COST_LIVING";
QUIT;
```

We get the following output, which is a snapshot of the Work library:

Library Name	Member Name	Member Type	DBMS Member Type	Data Set Label	Data Set Type	Date Created	Date Modified	Number of Physical Observations
WORK	COST_LIVING	DATA			DATA	01SEP19:09:08:23	01SEP19:09:08:23	19

Library Name	Member Name	Member Type	Column Name	Column Type	Column Length	Column Position	Column Number in Table	Column Label	Column Format
WORK	COST_LIVING	DATA	City	char	12	80	1		
WORK	COST_LIVING	DATA	Index	num	8	0	2		
WORK	COST_LIVING	DATA	Prev_yr_index	num	8	8	3		
WORK	COST_LIVING	DATA	Housing	num	8	16	4		
WORK	COST_LIVING	DATA	Food	num	8	24	5		
WORK	COST_LIVING	DATA	Travel	num	8	32	6		
WORK	COST_LIVING	DATA	Utility	num	8	40	7		
WORK	COST_LIVING	DATA	Education	num	8	48	8		
WORK	COST_LIVING	DATA	Leisure	num	8	56	9		
WORK	COST_LIVING	DATA	Other	num	8	64	10		
WORK	COST_LIVING	DATA	Updated	num	8	72	11		DATE9.

The column information provides a lot of details about the variable type, length, the column number in the table, formats, index, and so on.

The following are the most commonly used dictionary tables and their purpose. For a full list of dictionary tables and views and their use, please refer to the SAS installation documentation:

Dictionary table	Purpose
COLUMNS	Information related to columns.
FORMATS	Lists all the formats defined.
INDEXES	Informs whether indexes have been created.
LIBNAMES	Lists all the library names. Usually, this is the first port of call if you are unaware of the various libraries.
MACROS	Lists any defined macros.
OPTIONS	All SAS system options in place are listed.
TABLES	Only information on currently defined tables is provided.
TITLES	Currently defined titles and footnotes are listed.
VIEWS	Currently defined data views are listed.

Role of _ALL_ and _IN_

The _ALL_ functionality can be used for notes in the SAS log or to call all the variables that have been specified in the data step. Let's illustrate the first use by modifying the dataset we used while understanding the usage of WHERE and IF:

```
Data Updated_2019;
Set Cost_Living;
Where Year(Updated) = 2019;
If Index >= 80;
Put _All_;
Run;
```

This will give us the following output:

```
City=Mumbai Index=83 Prev_yr_index=85 Housing=40 Food=10 Travel=15
Utility=15 Education=10 Leisure=9 Other=1 Updated=02JAN2019
 _ERROR_=0 _N_=3
 City=New York Index=89 Prev_yr_index=85 Housing=40 Food=10
Travel=15 Utility=10 Education=20 Leisure=5 Other=5 Updated=02JAN2019
 _ERROR_=0 _N_=5
 City=Tokyo Index=87 Prev_yr_index=85 Housing=40 Food=15 Travel=10
Utility=5 Education=15 Leisure=14 Other=1 Updated=02JAN2019
```

```
_ERROR_=0  _N_=10
NOTE: There were 11 observations read from the data set WORK.COST_LIVING.
      WHERE YEAR(Updated)=2019;
NOTE: The data set WORK.UPDATED_2019 has 3 observations and 11 variables.
```

We can see that this particular usage of _All_ has led to all the variables in the dataset to be listed.

Use the following code block to list the variable names in the PROC PRINT data step; we just asked for all the variables to be printed using _ALL_:

```
PROC PRINT DATA = Updated_2019;
VAR Updated;
RUN;

Title '_ALL_ in a Data Step';
PROC PRINT DATA = Updated_2019;
VAR _ALL_;
RUN;
```

This will give us the following output:

Obs	Updated
1	01JAN2019
2	01JAN2019
3	01JAN2019

ALL in a Data Step

Obs	City	Index	Prev_yr_index	Housing	Food	Travel	Utility	Education	Leisure	Other	Updated
1	Mumbai	83	85	40	10	15	15	10	9	1	01JAN2019
2	New York	89	85	40	10	15	10	20	5	5	01JAN2019
3	Tokyo	87	85	40	15	10	5	15	14	1	01JAN2019

As we know, _N_ is initially set to 1. Each time the DATA step loops past the DATA statement, the _N_ variable increments by 1. The value of _N_ represents the number of times the DATA step has iterated. Let's look at selecting observations based on _N_:

```
Data Test;
Set Updated_2019;
If 1 < _N_ <10;
Run;
```

This will give us the following output:

Obs	City	Index	Prev_yr_index	Housing	Food	Travel	Utility	Education	Leisure	Other	Updated
1	New York	89	85	40	10	15	10	20	5	5	01JAN2019
2	Tokyo	87	85	40	15	10	5	15	14	1	01JAN2019

Using the _N_ automatic counter, we have been able to exclude the first observation and print only the remaining observations. If you compare this with the earlier output of the table, you will see that the observation where the city is Mumbai has not been output.

But be careful when using _N_. I will demonstrate this by trying to create a variable based on _N_:

```
Data Automatic;
Input A $ B;
Counter = _N_;
Datalines;
X 1
Y 2
Z 3
;
Run;

Data Automatic_Challenge;
Input A $ B;
Retain Counter 2;
_N_ = Counter+1;
Test_N = _N_;
Datalines;
X 1
Y 2
Z 3
;
Run;
```

This will give us the following output:

_N_ automatic value			
Obs	A	B	Counter
1	X	1	1
2	Y	2	2
3	Z	3	3

_N_ automatic value overwritten				
Obs	A	B	Counter	Test_N
1	X	1	2	3
2	Y	2	2	3
3	Z	3	2	3

In the first program, we can see that the automatic variable _N_ has a value between 1 and 3 and this corresponds to the observation count. In the second program, we have used the RETAIN function to assign the value 2 to the variable counter. We have created a derived variable called _N_. When we assign a new variable and try to point it to the automatic variable _N_, we get the value of the derived variable. We have, in this instance, confused SAS about which _N_ we are referring to, that is, the automatic variable or the derived one. Please don't assume that just because _N_ is an automatic variable if you use _N_ without caution, then SAS will still understand that you want the automatic variable's value to be output.

Summary

In this chapter, we covered the basics of SAS dataset fundamentals and how to create a table, compile it, and then execute it. We also looked into the basic syntax of the SAS programming language. Then, we learned how to compress, encrypt, and index a dataset. Finally, we learned the various operators in SAS, and how to format and subset the dataset.

In the next chapter, we will learn how to manipulate, sanitize, and transform data.

2
Data Manipulation and Transformation

This chapter will focus on introducing SAS capabilities that can help us to manipulate and transform data effectively. This chapter builds on some of the functions that we introduced in Chapter 1, *Introduction to SAS Programming*. The functions shared in this chapter will focus on capabilities that are tactical in nature and help to perform the task at hand. Some of the functions are similar in nature to what users of Microsoft Excel would use. However, the breadth and scope of these functions are expansive in SAS.

The number of functions in each SAS software update has been increasing. In the 9.4 release, there are more than 500 functions. These functions can be categorized as follows:

- Character string matching and manipulation (including concatenation, replacement, and truncation)
- Currency conversion
- Date and time
- Descriptive statistics
- Financial ratios
- Logic and control
- Mathematical and trigonometric functions
- Probability-related
- Macros
- Systems-related (SAS file I/O, external file support and routines, web tools, and so on)

While it is impossible to deal with all of these functions in this book, this chapter will introduce you to some of the most useful ones. We will continue to explore more functions in the following chapters.

Along with introducing more functions, we will also have a look at the LOOP functionality and the concept of BY GROUP processing to leverage the `FIRST` and `LAST` variables.

In this chapter, we will be looking at some of the variable manipulation tasks that data analysts frequently execute. The following are the major topics that will be covered:

- Length of a variable
- Case conversion and alignment
- String identification
- Dealing with blanks
- Missing and multiple values
- Interval calculations
- Concatenation
- Number conversion

Length of a variable

Prior to attempting any character manipulation, an important aspect to understand is how length is assigned for character variables. Let's look at the following character variable default length example:

```
DATA Cars;
INPUT Make $;
DATALINES;
Porsche_Cayenne
Audi
BMW
;
PROC CONTENTS;
RUN;
```

Using PROC CONTENTS, we get the following output:

The CONTENTS Procedure				
Data Set Name	WORK.CARS		Observations	3
Member Type	DATA		Variables	1
Engine	V9		Indexes	0
Created	09/01/2019 14:08:39		Observation Length	8
Last Modified	09/01/2019 14:08:39		Deleted Observations	0
Protection			Compressed	NO
Data Set Type			Sorted	NO
Label				
Data Representation	SOLARIS_X86_64, LINUX_X86_64, ALPHA_TRU64, LINUX_IA64			
Encoding	utf-8 Unicode (UTF-8)			

Alphabetic List of Variables and Attributes			
#	Variable	Type	Len
1	Make	Char	8

Even though, in the first observation, we have has 15 characters, the output of the first observation will get restricted to Porsche_ (the first eight characters) due to the default length of 8 for the character variable. In Chapter 1, *Introduction to SAS Programming*, we specified the length of the City variable, by using the $12. format in the input statement. In the case of the preceding program, no length has been specified. We have, however, specified a character format using the $ option. Please ensure that the length of the character variable is specified; otherwise, unintended consequences may occur in data processing and analysis.

In the following code block, we will use multiple methods to specify the length of the character variable:

```
Data Cars;
Length Make $ 15. Default=4;
Input Make $ Year;
Datalines;
Porsche_Cayenne 2018
Audi 2016
BMW 2014
;
```

We have now specified the desired length of the Make variable. In addition to that, we have introduced a numeric variable, Year, and specified its length using a different argument than what we used in Chapter 1, *Introduction to SAS Programming*. We can only specify the length of a numeric variable using the DEFAULT option.

Using PROC PRINT and CONTENTS, we get the following outputs specifying the length prior to Datalines:

Alphabetic List of Variables and Attributes			
#	Variable	Type	Len
1	Make	Char	15
2	Year	Num	4

Obs	Make	Year
1	Porsche_Cayenne	2018
2	Audi	2016
3	BMW	2014

You shouldn't be confused between the LENGTH format and function. In this chapter, we have, until now, specified the length format. However, there is also a LENGTH function that is available to SAS users. We will look at the LENGTH and LENGTHc functions while also reevaluating the format specification:

```
Data Cars;
Length¹ Make $ 5.;
Input Make $;
Datalines;
Audi
;
Data Length²;
Set Cars;
Length_Trimmed=Length³(Make);
Length_Non_Trimmed=Lengthc⁴(Make);
Run;
```

The following outputs show the varying uses of LENGTH:

Data Set Name	WORK.LENGTH			Observations	1
Member Type	DATA			Variables	3
Engine	V9			Indexes	0
Created	09/01/2019 14:18:31			Observation Length	24
Last Modified	09/01/2019 14:18:31			Deleted Observations	0
Protection				Compressed	NO
Data Set Type				Sorted	NO
Label					
Data Representation	SOLARIS_X86_64, LINUX_X86_64, ALPHA_TRU64, LINUX_IA64				
Encoding	utf-8 Unicode (UTF-8)				

Alphabetic List of Variables and Attributes			
#	Variable	Type	Len
3	Length_Non_Trimmed	Num	8
2	Length_Trimmed	Num	8
1	Make	Char	5

Obs	Make	Length_Trimmed	Length_Non_Trimmed
1	Audi	4	5

We used LENGTH on four different occasions in the preceding code:

- The first occasion was to specify the format of the Make character variable using the LENGTH format.
- The second use was to name a dataset: The dataset name could have been any different, but we named it Cars to highlight how the user shouldn't confuse different uses of the word in a SAS program.
- In the third instance, we used the LENGTH function to find the number of characters of Make. The count of the characters does not include any leading or trailing blanks.
- The LENGTHc function retains the leading or trailing blanks and returns the value of 5 for the newly created variable, Length_Non_Trimmed.

You would have noticed that the variables that were created, that is, Length_Trimmed and Length_Non_Trimmed, both have a default length of 8 in the LENGTH dataset.

Case conversion and alignment

In this section, we will learn how to identify and change case. We will also look at case alignment functions.

LowCase, PropCase, and UpCase

Data needs to be formatted for presentation and this is where cell alignment and case conversion come in handy. First, we will look at lower, upper, and proper case conversion:

```
Data Case;
Set Cars;
Upper=UpCase(Make);
Proper=PropCase(Make);
Lower=LowCase(Upper);
Run;
```

This will give us the following output, showing the case conversion:

Obs	Make	Year	Upper	Proper	Lower
1	Porsche_Cayenne	2018	PORSCHE_CAYENNE	Porsche_cayenne	porsche_cayenne
2	Audi	2016	AUDI	Audi	audi
3	BMW	2014	BMW	Bmw	bmw

We have used the existing `Cars` dataset and created three new variables to showcase the use of case conversion. When we separate two words using an underscore or some other special character, SAS still treats it as a single string. Hence, the first observation does not have the word Cayenne in a proper case.

However, the flexibility offered in SAS allows us to write a second argument to the `PropCase` function and overcome this problem. Add the following line of code to the `Case` dataset:

```
Proper_second_argument=PropCase(Make, "_");
```

You will get the following output, showing the case conversion of the second word:

Obs	Make	Year	Upper	Proper	Lower	Proper_second_argument
1	Porsche_Cayenne	2018	PORSCHE_CAYENNE	Porsche_cayenne	porsche_cayenne	Porsche_Cayenne
2	Audi	2016	AUDI	Audi	audi	Audi
3	BMW	2014	BMW	Bmw	bmw	Bmw

The second argument helps to ignore the special characters specified.

Up until now, we have seen how to convert text into the desired case. What if you need to find out whether a particular case format has been used on existing data? For this, we can use the AnyUpper, AnyLower, and NoTupper functions, which you will learn about in the following section.

AnyUpper, AnyLower, and NoTupper

As their names suggest, the first two functions look for any upper- and lowercase characters, respectively. The third function checks whether any character in the string is not in uppercase. All three of them return the position of the first character that matches the condition.

With all three functions, you can specify a start position as an argument. The rules for the start position are as follows (http://support.sas.com/documentation/onlinedoc/91pdf/sasdoc_913/base_lrdictionary_10307.pdf):

- If the value of the starting position is positive, the search proceeds to the right
- If the value of the starting position is negative, the search proceeds to the left
- If the value of the starting position is less than the negative length of the string, the search begins at the end of the string

The functions return a value of zero when either of the following occurs:

- The character that you are searching for is not found
- The value of the starting position is greater than the length of the string
- The value of start equals zero

Let's try and test the functions on the Cars dataset in the following code block:

```
Data Case_Test;
Set Cars;
Upper_Pos = AnyUpper(Make);
Lower_Pos = AnyLower(Make);
Tupper_Pos = NoTupper(Make);
Run;
```

The following output shows the number of characters identified in each case:

Obs	Make	Year	Upper_Pos	Lower_Pos	Tupper_Pos
1	Porsche_Cayenne	2018	1	2	2
2	Audi	2016	1	2	2
3	BMW	2014	1	0	4

The Upper_Pos variable has a value of 1 as the AnyUpper function has found that the first letter is in uppercase in all three observations. The Lower_Pos variable has the value 2 for the first observation as the second letter is in lowercase. In the third observation, the variable has a value of 0 since there are no lowercase letters. The NoTupper function worked just like AnyLower and produced the same values for the first two observations of Tupper_Pos just as in the values for Lower_Pos. However, for the third observation, Tupper_Pos has a value of 4. Even though no lowercase was found, the position of the end of the characters in the Make variable was output.

Left and right

To align the data, let's look at the basic Left and Right functions:

```
Data Align;
Set Cars;
Char_right=Right(Make);
Num_left=Left(Year);
Run;
```

The following output shows the aligned data:

Obs	Make	Year	Char_right	Num_left
1	Porsche_Cayenne	2018	Porsche_Cayenne	2018
2	Audi	2016	Audi	2016
3	BMW	2014	BMW	2014

Character variables are left-aligned and numeric variables are right-aligned by default. However, we created two new variables to make the character variable right-aligned and the numeric variable left-aligned. Alignment functions perform differently on the dataset, **Output Delivery System (ODS)**, and title statements.

String identification

In this section, we will explore the `Scan`, `Index`, and `Find` functions that help us with string identification.

The Scan function

Unless you can successfully identify a part of the string successfully, replacement or modification of the string value is impossible. In the following code block, we will review some functions that help in string identification:

```
Data Cities;
Input City $50.;
First=Scan(City, +1);
Last=Scan(City,-1);
Third=Scan(City,4);
Datalines;
Chicago Paris London Geneva Dublin
;
```

The following output shows the words identified from the given string:

Obs	City	First	Last	Fourth
1	Chicago Paris London Geneva Dublin	Chicago	Dublin	Geneva

The `First` and the `Last` words in the string are identified using the positive and negative count. The positive count looks for the first instance of a word in a string from left to right. A negative count looks for the word from right to left. To find the fourth instance of a word in a string, we simply specify 4 in the argument. This is a simplistic example where no delimiter, apart from space, is present in the string.

To further showcase the benefit of the `Scan` function, the loop facility in SAS will be used. While we will discuss loops and macros in detail in `Chapter 5`, *Advanced Programming Techniques – SAS Macros*, loops are discussed in this chapter briefly.

In case you haven't leveraged loops already, they are an effective way to deal with repetitive tasks. Loops can also help to perform iterative tasks based on a rule for multiple runs of a program. We will explore the case of a 0% interest loan that needs to be paid off. The total amount of the loan is $26,000 and the first payment of $2,000 has already been made. Let's see what the amount outstanding will be after one year if a monthly payment of $2,000 is made:

```
Data Two_Year_Payment;
Initial = 26000;
Balance = Initial - 2000;
do i = 1 to 12;
Balance = Balance - 2000;
Output;
End;
Run;
```

The following output shows the outstanding amount after one year:

Obs	Initial	Balance	i
1	26000	22000	1
2	26000	20000	2
3	26000	18000	3
4	26000	16000	4
5	26000	14000	5
6	26000	12000	6
7	26000	10000	7
8	26000	8000	8
9	26000	6000	9
10	26000	4000	10
11	26000	2000	11
12	26000	0	12

We started with a borrowed amount of $26,000 and called the variable, `Initial` variable. The balance outstanding prior to the monthly payments was $26,000 – $2,000. The loop was initiated by the `do` argument, where we asked the loop to run from the 1st to the 12th iteration. The `Balance = Balance - 2000` argument specified the iteration that we wanted each instance of the loop to perform. We specified the output command as we needed to see the outcome for all the iterations of the loop. Without the output command, we would only get the result after the last iteration has been run.

Every loop needs an end command, just as it needs a start command using the do argument. From the results, we can see that we will pay off the borrowed amount after making 12 payments of $2,000 each.

 We will be using the concept of loops to showcase the use of other SAS functions. Aspects such as loop, until, while, and nested loops will be discussed later in this book.

Let's now use the Scan function in a loop to identify words in a string and break them down into multiple variables. This is possible to do without loops, but it's much easier to complete repetitive tasks in the following manner:

```
Data Scan_in_loop;
Length thedebate $50;
thedebate = "Is Pluto a Planet, well yes, and no";
delim = ',';
modif = 'oq';
nwords = CountW(thedebate, delim, modif);
Do count = 1 to nwords;
words = scan(thedebate, count, delim, modif);
Output;
End;
Run;
Proc Print;
Run;
```

The following is the resultant output:

Obs	thedebate	delim	modif	nwords	count	words
1	Is Pluto a Planet, well yes, and no	,	oq	3	1	Is Pluto a Planet
2	Is Pluto a Planet, well yes, and no	,	oq	3	2	well yes
3	Is Pluto a Planet, well yes, and no	,	oq	3	3	and no

Among some of the new things you have been introduced to in the preceding code, the delimiter and modifier options are quite powerful.

 A delimiter is any of several characters that are used to separate words. You can specify the delimiters in the charlist and modifier arguments.

If you specify the Q modifier, then delimiters inside of substrings that are enclosed in quotation marks are ignored. Since we have specified the Q modifier without the M modifier, the Scan function does the following documentation (http://support.sas.com/ documentation/cdl/en/mcrolref/62978/PDF/default/mcrolref.pdf):

- Ignores delimiters at the beginning or end of the string
- Treats two or more consecutive delimiters as if they were a single delimiter

If the string contains no characters other than delimiters, or if you specify a count that is greater in absolute value than the number of words in the string, then the SCAN function returns one of the following:

- A single blank when you call the SCAN function from a DATA step
- A string with a length of zero when you call the SCAN function from the macro processor (http://support.sas.com/documentation/cdl/en/lrdict/64316/ PDF/default/lrdict.pdf).

In the SCAN function, the word refers to a substring that has all of the following characteristics:

- It's bounded on the left by a delimiter or the beginning of the string.
- It's bounded on the right by a delimiter or the end of the string.
- It doesn't contain delimiters.

A word can have a length of zero if there are delimiters at the beginning or end of the string, or if the string contains two or more consecutive delimiters. However, the SCAN function ignores words that have a length of zero unless you specify the M modifier.

Apart from the Q modifier, we have also used the O modifier. The O modifier processes the charlist and modifier arguments once, rather than every time the SCAN function is called. Using the O modifier in the DATA step (excluding WHERE clauses) or in the SQL procedure can make SCAN run faster when you call it in a loop where the charlist and modifier arguments do not change. The O modifier applies separately to each instance of the SCAN function in your SAS code and does not cause all the instances of the SCAN function to use the same delimiters and modifiers.

Knowledge of modifiers is crucial as they can be used as arguments in other functions, such as COMPRESS.

The other modifiers are as follows (they aren't case-sensitive) and can be found on the official SAS documentation (http://support.sas.com/documentation/cdl/en/lefunctionsref/63354/HTML/default/viewer.htm#p0jshdjy2z9zdzn1h7k90u99lyq6.htm):

Modifier	Feature
A	This adds alphabetic characters to the list of characters.
B	This scans backward from right to left instead of from left to right, regardless of the sign of the count argument.
C	This adds control characters to the list of characters.
D	This adds digits to the list of characters.
F	This adds an underscore and English letters (that is, valid first characters in a SAS variable name using VALIDVARNAME=V7) to the list of characters.
G	This adds graphic characters to the list of characters. Graphic characters are characters that, when printed, produce an image on paper.
H	This adds a horizontal tab to the list of characters.
I	This ignores the case of the characters.
K	This causes all characters that are not in the list of characters to be treated as delimiters. That is, if K is specified, then characters that are in the list of characters are kept in the returned value rather than being omitted because they are delimiters. If K is not specified, then all the characters that are in the list of characters are treated as delimiters.
L	This adds lowercase letters to the list of characters.
M	This specifies that multiple consecutive delimiters and delimiters at the beginning or end of the string argument refer to words that have a length of zero. If the M modifier is not specified, then multiple consecutive delimiters are treated as one delimiter, and delimiters at the beginning or end of the string argument are ignored.
N	This adds digits, an underscore, and English letters (that is, the characters that can appear in a SAS variable name using VALIDVARNAME=V7) to the list of characters.
P	This adds punctuation marks to the list of characters.
R	This removes leading and trailing blanks from the word that SCAN returns.
S	This adds space characters to the list of characters (blank, horizontal tab, vertical tab, carriage return, line feed, and form feed).
T	This trims trailing blanks from the string and charlist arguments.

U	This adds uppercase letters to the list of characters.
W	This adds printable (writable) characters to the list of characters.
X	This adds hexadecimal characters to the list of characters.

If the modifier argument is a character constant, then enclose it in quotation marks. Specify multiple modifiers in a single set of quotation marks. A modifier argument can also be expressed as a character variable or expression.

In our query, the `CountW` function counts the number of words in a character string. The loop will run three times, which is equivalent to the number of words in the character string based on the delimiter and modifier we supplied as arguments. The delimiter comma has been used to identify phrases in the `thedebate` variable. Words prior to each instance of the comma have been treated as a separate variable. As a result, we end up with three variables in the output.

Index, Indexc, and Indexw

Another set of useful functions for finding information in a string are the variations of the `Index` function. Let's look at the following example to understand how it works:

```
Data Index_Demo;
    String_x = 'Indexit(FINDIT)';
    String_y = 'findit';
    Not_Found = Index(String_x, String_y);
    Found=Index(String_x,upcase(String_y));
Run;
```

The following is the resultant output:

Obs	String_x	String_y	Not_Found	Found
1	Indexit(FINDIT)	findit	0	9

The `Index` function searches for the first occurrence of the character string as a substring. In this example, we are trying to find the starting position of the `FINDIT` string within the `String_x` variable. We used the `String_x` variable as the first argument in the `Index` function.

The second argument, which is represented by the String_y variable, consists of the string that we are trying to find in String_x. In the first attempt, for the Non_Found variable, we get a value of 0. However, for the same string, in the second attempt, we get a value of 9 for the Found variable. Please note that the only difference between both attempts is the case difference for FINDIT. Since the first argument has it in uppercase, we used the upcase function to find the first instance of the string.

When trying to find strings within a variable, you may need to leverage multiple functions at the same time to get to the right answer.

The Indexc function provides more flexibility than Index. The Indexc function searches the source, from left to right, for the first occurrence of any character present in the excerpts and returns the position in the source of that character. If none of the characters in excerpt-1 through excerpt-n in the source are found, Indexc returns a value of 0:

```
Data Indexc;
String="It's confusing";
Answer_A=Indexc(String,'sortit','how?');
Answer_B=Indexc(String,'sortit',"i can't");
Answer_C=Indexc(String,'sortit','I can');
Run;
```

The following is the resultant output:

Obs	String	Answer_A	Answer_B	Answer_C
1	It's confusing	2	2	1

Instead of two arguments that we had for the Index function, we now have three arguments for Indexc. While creating the Answer_A variable, we are trying to find the position of the character in the String variable, where the character matches the ones that were either the second or third argument. In the first instance of using the function, we get the answer 2. This is because the letters s, o, t, and i from the first argument and the letter o from the second argument are found in the String variable. Out of s, o, t, and i, the first letter that matches exactly (including the case) is t. It doesn't matter that the first letter of sortit matches the first argument. Indexc is going to provide the answer on the basis of which is the first character while scanning from left to right to ensure that matches with the second or third argument.

The role of cases in matching characters is important and we have already seen a few examples of it. This is further highlighted in the answer we get in the second and third incidence of using Indexc. In the third instance, the letter i matches the case in the first and third arguments. Since that is the first letter in the first argument, we get the answer 1.

Find

Another powerful function to help us search strings is the `Find` function. The `Find` function searches strings for the first occurrence of the specified substring and returns the position of that substring. If the substring is not found in string, `Find` returns a value of 0.

If `startpos` is not specified, `Find` starts the search at the beginning of the string and searches the string from left to right. If `startpos` is specified, the absolute value of `startpos` determines the position at which to start the search. The sign of `startpos` determines the direction of the search. The `Find` function and the `Index` function both search for substrings of characters in a character string. However, the `Index` function does not have the modifiers, nor the `startpos` arguments. We will look at an example in the following code block:

```
Data Find;
String = "We will explore the FIND function. Won't we?";
String_Length = Length(String);
Answer = Find(String, "we");
Run;
```

This will result in the following output:

Obs	String	String_Length	Answer
1	We will explore the FIND function. Won't we?	44	42

The `Find` function is case-sensitive and ignores the first instance of we in proper case. To make it case-insensitive, let's leverage the modifiers that we discovered earlier in this chapter, as shown in the following code block:

```
Data Non_Case;
String = "We will explore the FIND function. Won't we?";
String_Length = Length(String);
Answer = Find(String, "we", "i");
Run;
```

This will result in the following output:

Obs	String	String_Length	Answer
1	We will explore the FIND function. Won't we?	44	1

The i modifier has helped us ignore the case and hence our answer is the first position of the string.

Now, let's use the `startpos` functionality that the `Find` function offers:

```
Data Startpos;
String = "We will explore the FIND function. Won't we?";
Startposvar = 2;
String_Length = Length(String);
Answer = Find(String, "we", "i", Startposvar);
Run;
```

This will result in the following output:

Obs	String	Startposvar	String_Length	Answer
1	We will explore the FIND function. Won't we?	2	44	42

Since we have specified the start position as 2, we ignore the first instance of we even though there is a case insensitive argument specified.

We will now look at another example with a negative position, as shown in the following code block:

```
Data Negativestart;
String = "We will explore the FIND function. Won't we?";
Startposvar = 3-44;
String_Length = Length(String);
Answer = Find(String, "we", "i", Startposvar);
Run;
```

This will result in the following output:

Obs	String	Startposvar	String_Length	Answer
1	We will explore the FIND function. Won't we?	-41	44	1

This time, we specified a negative position to start searching the string. Remember, if the start position is greater than 0, the `Find` function starts the search at the `startpos` position and the direction of the search is to the right. If `startpos` is greater than the length of the string, `Find` returns a value of 0. If the start position is less than 0, it searches at the `startpos` position and the direction of the search is to the left. If `startpos` is greater than the length of the string, the search starts at the end of the string. If the value of `startpos` is 0, then it returns the value 0.

We will now look at an example of using leading and trailing with `Find`:

```
Data Leadingandtrailing;
String = "We will explore the FIND function. Won't we?";
Startposvar = 1;
String_Length = Length(String);
Answer = Find(String, "   explore   ", "i", Startposvar);
Answer_1 = Find(String, "explore ", "t", Startposvar);
Answer_2 = Find(String, " explore ", "t", Startposvar);
Run;
```

This will result in the following output:

Obs	String	Startposvar	String_Length	Answer	Answer_1	Answer_2
1	We will explore the FIND function. Won't we?	1	44	0	9	8

The answer variable gives a value of 0 as the modifier used in the preceding example is incorrect for the data problem. We need to use the t modifier, which helps us deal with leading and trailing blanks. `answer_2` showcases how, with the use of the modifier, the starting position changes to 8 (instead of 9 for `answer_1`), pointing to the fact that the blank has been accounted for in the FIND function.

> You cannot specify all of the modifiers in an argument. For instance, the FIND function accepts up to four modifiers. In an ideal data process, never specify non-essential modifiers as it can lead to unforeseen consequences. This can happen easily if you are dealing with a large amount of data.

Dealing with blanks

Blanks, either system-generated or as part of the source data, can cause problems when retained in a string. The following functions help deal with them.

Compress, Strip, and Trim

The compress, strip, and trim functions help us deal with blanks in strings. Based on the following example, try and figure out the differences in the Compress, Strip, and Trim functions:

```
Data Compare;
Length String $20.;
```

```
Format String $20.;
String = " 3 fn comparison ";
Compress="#"||Compress(String)||"#";
Trim="#"||Trim(String)||"#";
Strip="#"||Strip(String)||"#";
Run;
```

This will result in the following output:

Obs	String	Compress	Trim	Strip
1	3 fn comparison	#3fncomparison#	# 3 fn comparison#	#3 fn comparison#

We have used the concatenation (||) parameters to highlight how these three functions deal with blanks.

COMPRESS removes all blanks from the string (not just the leading and trailing). TRIM only removes the trailing blank whereas STRIP has removed the leading and trailing blanks.

Missing and multiple values

There are instances in databases where the same variables values are stored in multiple tables or systems. For instance, the same employment details field captured for a customer may be missing or may have different values across the various contacts points that it may have been collected. The value might be accurate and updated in a credit application but may not be updated in a social media profile. The functions in the following section help to deal with missing and multiple values.

COALESCE and COALESCEC

If the objective is to only select the first non-missing value from amongst the options, we can use COALESCE for numeric and the COALESCEC function for character values.

Both functions accept multiple arguments. They check the value of each argument in the order in which they are specified and output the first non-missing value. If only one value is present, then the functions return that value. If none of the arguments have a value, then the function returns a missing value (not 0). We will see how these two functions work in the following example:

```
Data Select;
A = Coalesce(1,2,3,4,5,6,7,8,9,10,1);
B = Coalesce(1,2,.,.,5,6,7,8,9,10,1);
C = Coalesce(1.,2.,.,.,5,6,7,8,9,10,1);
D = Coalesce(.,.,.,.,.,.,.,.,.,.,.);
E = Coalescec(0,2,.,.,5,6,7,8,9,10,1.);
F = Coalescec(0,2,.,.,5,6,7,8,9,10,'1#', 'Choose me');
G = Coalescec('Choose me', 'No!');
Run;
```

This will result in the following output:

Obs	A	B	C	D	E	F	G
1	1	1	1	.	0	0	Choose me

For the A, B, and C variables, we get the answer as 1 due to the fact that the first variable in the COALESCE function is non-missing and has a value of 1. It doesn't matter what the values of the rest of the arguments are in the function. The function will output the first non-missing argument. In the case of the D variable, we don't have any values in the variable and hence the output is a missing value.

The E variable looks numeric but it has been created as Char as SAS has automatically assigned this attribute owing to the last argument presented. Hence, the COALESEC function does not produce an error and gives the output as 0. Similarly, for the F and G variables, we get the first character argument as the output.

Interval calculations

Variables containing dates will often be used to find the time for an event. To find the answer, interval calculations will come in handy. We will explore the INTNX and INTCK functions in the following section.

INTNX and INTCK

The most useful functions for calculating date values, datetime values, and time intervals are INTNX and INTCK. These functions help in interval calculations. In common coding parlance, you can refer to INTNX as an interval check and INTCK as an interval next function. What this means is that INTNX checks for intervals whereas INTCK is useful for computing a date/datetime value on the basis of a different date/datetime value. Apart from this difference, there is a minor difference in the syntax. The INTNX function has four arguments, whereas the INTCK function has three arguments. The first two arguments are the same for both functions:

- interval: A character constant or variable that contains an interval name
- from: An SAS data/datetime value

The remaining arguments for INTNX are as follows:

- n: The number of intervals to increment from the interval that contains the from value
- alignment: Controls the alignment of dates, with the allowed values being BEGINNING, MIDDLE, END, and SAMEDAY

The remaining argument for INTCK is as follows:

- to: Ending date/datetime value

The forms of both functions are as follows:

```
INTNX (interval, form, n <alignment>);
INTCK (interval, form, to);
```

Let's look at some examples of INTNX:

```
Data Emissions;
Input Year Month $3. Coal Gas Petrol Diesel Nuclear;
Datalines;
2018 Jan 110 112 113 114 112
2018 Feb 110 113 114 116 112
2018 Mar 112 114 114 116 110
2018 Apr 114 115 113 115 111
2018 May 116 114 112 114 110
;
Data Add_Month;
Set Emissions;
Format Date_next Date9.;
Date_next = INTNX ('Month', '1Jan2018'd, _n_);
```

```
Format Date_current Date9.;
Date_current = INTNX ('Month', '1Jan2018'd, _n_ - 1);
Format Date_plus_one Date9.;
Date_plus_one = INTNX ('Month', '1Jan2018'd, _n_ + 1);
Run;
```

This will result in the following output:

Obs	Year	Month	Coal	Gas	Petrol	Diesel	Nuclear	Date_next	Date_current	Date_plus_one
1	2018	Jan	110	112	113	114	112	01FEB2018	01JAN2018	01MAR2018
2	2018	Feb	110	113	114	116	112	01MAR2018	01FEB2018	01APR2018
3	2018	Mar	112	114	114	116	110	01APR2018	01MAR2018	01MAY2018
4	2018	Apr	114	115	113	115	111	01MAY2018	01APR2018	01JUN2018
5	2018	May	116	114	112	114	110	01JUN2018	01MAY2018	01JUL2018

In this example, we have assumed that the report date generation on change in emission index values is generated on the 1st of each month. We have tried to introduce a variable with the monthly date. Three different variables are generated to showcase that the number of increments to n can be zero, positive, or negative.

The following example will further illustrate how the INTNX function works:

```
Data Interval_Days;
Set Add_Month (Drop = Date_next Date_plus_one);
Interval_Days
= INTNX('Month', Date_Current, 1) - INTNX('Month', Date_Current, 0);
Interval_Days1
= INTNX('Month', Date_Current, 2) - INTNX('Month', Date_Current, 0);
Interval_Days2
= INTNX('Month', Date_Current, -1) - INTNX('Month', Date_Current, 0);
Interval_Days3
= INTNX('Month', Date_Current, 3) - INTNX('Month', Date_Current, 0);
Run;
```

This will result in the following output:

Obs	Year	Month	Coal	Gas	Petrol	Diesel	Nuclear	Date_current	Interval_Days	Interval_Days1	Interval_Days2	Interval_Days3
1	2018	Jan	110	112	113	114	112	01JAN2018	31	59	-31	90
2	2018	Feb	110	113	114	116	112	01FEB2018	28	59	-31	89
3	2018	Mar	112	114	114	116	110	01MAR2018	31	61	-28	92
4	2018	Apr	114	115	113	115	111	01APR2018	30	61	-31	91
5	2018	May	116	114	112	114	110	01MAY2018	31	61	-30	92

While calculating the `Interval_days` variable, the first component (`INTNX('Month',  Date_Current, 1)`) resolves to `01Feb2018` and the second component (`INTNX('Month', Date_Current, 0)`) resolves to `01Jan2018` for the first observation on the `Date_Current` variable. Hence, we get `31` days as the answer. For the first component, one period further from `01Jan2018` has been calculated. The period has been specified as `Month`. When we specify the component as (`INTNX('Month', Date_Current, -1)`), we are asking the function to output the value one period before the value held by the `Date_Current` variable. Again, the period has been specified as a month.

We can also use the `INTNX` function to compute the ceiling of an interval, as shown in the following code block:

```
Data Cieling_Years;
Set Add_Month (Drop = Date_next Date_plus_one);
Format OldYear Year4.;
Format NewYear Year4.;
OldYear = INTNX('Year', Date_Current + 1, -1);
CurrentYear = Year(Date_Current);
NewYear = INTNX('Year', Date_Current, 1);
Run;
```

This will result in the following output:

Obs	Year	Month	Coal	Gas	Petrol	Diesel	Nuclear	Date_current	OldYear	NewYear	CurrentYear
1	2018	Jan	110	112	113	114	112	01JAN2018	2017	2019	2018
2	2018	Feb	110	113	114	116	112	01FEB2018	2017	2019	2018
3	2018	Mar	112	114	114	116	110	01MAR2018	2017	2019	2018
4	2018	Apr	114	115	113	115	111	01APR2018	2017	2019	2018
5	2018	May	116	114	112	114	110	01MAY2018	2017	2019	2018

We have been able to create variables that hold the values of the previous, current, and next year.

Up until now, we have not explored the `Alignment` argument while using the `INTNX` function. We will explore its uses in the following example:

```
Data Alignment;
Format Beginning Date9.;
Beginning = INTNX('Month', '31Jan2019'd, 7, 'Beginning');
Format Middle Date9.;
Middle = INTNX('Month', '31Jan2019'd, 7, 'Middle');
Format End_ Date9.;
End_ = INTNX('Month', '14Jan2019'd, 7, 'End');
Format SameDay Date9.;
Sameday = INTNX('Month', '31Jan2019'd, 7, 'Sameday');
Run;
```

This will result in the following output:

Obs	Beginning	Middle	End_	SameDay
1	01AUG2019	16AUG2019	31AUG2019	31AUG2019

The `Alignment` argument can also be specified using just the first letter of each argument. It's much easier to calculate the beginning of the end of each month. However, `Middle` and `Sameday` are more powerful options and save a good amount of coding steps that would be needed in their absence.

The following table will help us better understand the `INTCK` function with the help of SAS statements and their expected results:

SAS Statements	Results
INTCK ('Year', '01Jan2011'd, '01Aug2019'd);	8
INTCK ('Days365', '01Jan2011'd, '01Aug2019'd);	8
INTCK ('Year', '31Dec2018'd, '01Jan2019'd);	1
INTCK ('Days365', '31Dec2018'd, '01Jan2019'd);	0
INTCK ('Month', '01Jan2011'd, '01Aug2019'd);	103
INTCK ('Days', '01Jan2011'd, '01Aug2019'd);	3134
INTCK ('SemiYear', '01Jan2011'd, '01Aug2019'd);	17
INTCK ('Qtr', '01Jan2011'd, '01Aug2019'd);	34
INTCK ('Hour', '14:00:56't, '23:45:54't);	9
INTCK ('Minute', '14:00:56't, '23:45:54't);	585
INTCK ('Second', '14:00:56't, '23:45:54't);	35098

There is an optional argument called `method` for the `INTCK` function. It specifies whether a continuous or discrete method is used to count the intervals. Discrete is the default method that the function uses. You can either specify the continuous option as `C` or `CONT` and the discrete option as `D` or `DISC`. We will explore the `INTCK` function in the following example:

```
Data Method;
Input Type $ Production Jul :Date9. Aug :Date9.;
Format Jul :Date9. Aug :Date9.;
Datalines;
W/e 131 07Jul2019 06Oct2019
W/e 234 14Jul2019 13Oct2019
W/e 232 21Jul2019 20Oct2019
W/e 212 28Jul2019 27Oct2019
M/e 203 31Jul2019 31Oct2019
;
Data Comparison;
Set Method;
Month_D = INTCK ('Month', Jul, Aug);
Month_C = INTCK ('Month', Jul, Aug, 'C');
Run;
```

This will result in the following output:

Obs	Type	Production	Jul	Aug	Month_D	Month_C
1	W/e	131	07JUL2019	06OCT2019	3	2
2	W/e	234	14JUL2019	13OCT2019	3	2
3	W/e	232	21JUL2019	20OCT2019	3	2
4	W/e	212	28JUL2019	27OCT2019	3	2
5	M/e	203	31JUL2019	31OCT2019	3	3

The discrete and continuous methods provide us with different answers for all but the last observation. The discrete method calculates the months based on the calendar months involved, whereas the continuous method uses the exact dates. The only observation where they both yield the same result is for the last observation.

There are many intervals available in SAS to use with the `INTNX` and `INTCK` functions. You can devise your own custom intervals.

 Please see **SAS Studio Help** for further information.

There are three variations of the WEEK interval that have been used in the following code block to showcase how a slight difference in the interval options can help to generate customized results:

```
Data _Null_;
Format Week Date9. Week_Sun Date9. Week_Mon Date9.;
Week = INTNX ('Week', '01Jan2019'd+1, 3);
Week_Sun = INTNX ('Week2', '01Jan2019'd+1, 3);
Week_Mon = INTNX ('Week.2', '01Jan2019'd+1, 3);
PUT 'Week= ' Week;
PUT 'Week_Sun = ' Week_Sun;
PUT 'Week_Mon = ' Week_Mon;
Run;
```

This will result in the following message written to LOG:

```
Week= 20JAN2019
Week_Sun = 03FEB2019
Week_Mon = 21JAN2019
NOTE: DATA statement used (Total process time):
         real time              0.00 seconds
         cpu time               0.00 seconds
```

The preceding variable resolution should help you understand the effects of interval options on the output.

The WEEK interval helped us find the start of the third week from January 2. The WEEK2 interval calculates the intervals biweekly and hence produces a result of Feb 3, 2019. The WEEK.2 interval produces similar output to the WEEK interval except that it calculates the week intervals starting on Monday.

Hopefully, this has provided insight into how minor modifications to the interval will produce differing results.

Concatenation

When dealing with strings, multiple variables may need to be joined together with some form of concatenation function. We will explore some of these options in the following sections.

CAT

We explored COALESCE and related functions to help us choose from multiple values. However, there are other instances when we want to join observations together. Earlier, the concatenation operator (| |) was mentioned when we showcased the difference between the Compress, Strip, and Trim functions. The CAT function is a built-in function in SAS to help join observations. Let's compare it with the use of the concatenation operator in the following code block:

```
73          Data _Null_;
74          A = "This ";
75          B = " is";
76          C = " a test      ";
77          D = " of CAT function";
78          Out_Symbol = A||B||C||D;
79          Out_CAT = CAT (A, B, C, D);
80          Put Out_Symbol;
81          Put Out_CAT;
82          Run;
```

This will result in the following message in LOG:

```
This  is a test      of CAT function
This  is a test      of CAT function
NOTE: DATA statement used (Total process time):
      real time           0.00 seconds
      cpu time            0.00 seconds
```

None of the options we've used have dealt with leading and trailing blanks in the variables being joined. We would have got a different answer had we used a COMPRESS function with them. Let's also look at some of the variants of the CAT function that are available.

CATS, CATT, and CATX

Let's look at the use of the variations of the CAT functions:

```
Data Joins;
A = "This ";
B = " is";
C = " a test      ";
D = " of CAT";
Out_Symbol = Compress (A||B||C||D);
Out_CAT = Compress (CAT (A, B, C, D));
Out_CATS = CATS (A, B, C, D);
Out_CATT = CATT (A, B, C, D);
```

```
SP = '$';
Out_CATX = CATX (SP, A, B, C, D);
Run;
Proc Print Noobs;
Var Out_Symbol Out_CAT Out_CATS Out_CATT Out_CATX;
Run;
```

This will result in the following output:

Out_Symbol	Out_CAT	Out_CATS	Out_CATT	Out_CATX
ThisisatestofCAT	ThisisatestofCAT	Thisisa testof CAT	This is a test of CAT	Thisisa test$of CAT

If the goal is to remove all leading and trailing blanks in a string and output a sentence, the CATT function does the job for us. However, depending on the business problem, we can leverage different variations of the CAT function. The CATX function even allows us to embed a delimiter in the string.

The CATX function removes leading and trailing blanks, gives us the option of inserting delimiters, and concatenates the string, whereas the CAT function does not remove leading or trailing blanks but does concatenate the string. The CATT function does not remove leading banks. However, it does remove trailing blanks and concatenates the string. The CATS function does remove trailing and leading blanks and returns a concatenated string. The function doesn't support inserting delimiters such as CATX:

While the functions showcased in the preceding example produce different results, when combined with the use of TRIM, LEFT and the concatenation operator (| |), all these functions can produce similar functions. The reason to use the CAT family of functions is that they are faster than TRIM and LEFT. We can also combine the functions with the OF syntax to write the SAS statement in a slightly different way. The result remains the same:

CAT (OF A1-D1) = A1 | | A2 | | A3 | | A4

Here, A1, A2, A3, and A4 are four different variables.

Up until now, we have looked at an example of joining variables with non-missing values. We will now look at a missing value example:

```
Data Emissions_City;
Input UK $ US $ China $;
Datalines;
. Coal .
Gas Nuclear Petrol
Coal Gas .
```

```
. . Petrol
;
Data String_Missing;
Set Emissions_City;
SP = '"';
Delimiter = CATX (SP, UK, US, China);
Delimiter_Space = CATX ("", UK, US, China);
No_Delimiter = CATT (UK, US, China);
Run;
```

This will result in the following output:

Obs	UK	US	China	SP	Delimiter	Delimiter_Space	No_Delimiter
1		Coal		"	Coal	Coal	Coal
2	Gas	Nuclear	Petrol	"	Gas"Nuclear"Petrol	Gas Nuclear Petrol	GasNuclearPetrol
3	Coal	Gas		"	Coal"Gas	Coal Gas	CoalGas
4			Petrol	"	Petrol	Petrol	Petrol

Unlike the previous example, CATT doesn't do the job for us. Due to the missing observations, the output that's produced isn't right. In this example, the CATX function produces the right output. Whenever you are dealing with strings, the choice of the function should depend on what is presumed to be the right answer. The Delimiter variable might hold what some may call the right way to store the data in the variable.

LAGS

If you want to calculate a period of time between one event and another, then you should use the LAG function. This function helps to return values from a queue. The syntax has two elements:

- The first argument specifies a numeric or character constant, variable, or expression.
- The optional argument specifies the number of lagged values. The default is 1.

Let's look at lagging the variables:

```
Data Lag;
Set Decimal (Keep = Round);
Lag1 = Lag1 (Round);
Lag2 = Lag2 (Round);
Run;
```

This will result in the following output:

Obs	Round	Lag1	Lag2
1	67	.	.
2	53	67	.
3	45	53	67
4	61	45	53
5	80	61	45
6	69	80	61
7	69	69	80

By specifying a single lag, we have started the LAG1 variable values from the second observation. The first observation has been set to missing. In the case of LAG2, we have specified two lags. This is the simplest form of lagging. How can we create lags if we have scores of two different classes in the same dataset? This can be done by leveraging BY GROUP processing.

Before going further into discussing lags, it would be beneficial to explore the concept of the FIRST. and LAST. variables, which are automatically created with BY GROUP processing. Prior to using BY GROUP, we should ensure that the dataset has been sorted by the variable being referred to in BY GROUP.

The concept of FIRST. and LAST. can be explored by running the following code:

```
Data Lag_ByGroup;
Set Class_Scores;
By Class;
Lag1 = Lag1 (Score);
IF FIRST.Class THEN DO;
Lag1 = .;
END;
ELSE DO;
Lag1 = Lag1;
END;
RUN;
```

This will result in the following output:

Obs	Class	Score	Lag1
1	A	21	.
2	A	23	21
3	A	25	23
4	A	27	25
5	B	15	.
6	B	20	15
7	B	25	20
8	B	30	25

We have lagged the SCORE variable by one observation for each instance of the CLASS variable. Moreover, we have used the IF-THEN conditions to set the first of each observation as missing. If you want, you can now find the difference in the observation from its subsequent observation by finding the difference between LAG1 and SCORE. You can create the preceding dataset in multiple ways using different SAS functions.

Logic and control

We will look at three different sets of functions that will help us deal with logic and control issues.

IFC and IFN

There are also specific functions that return a value based on a condition that mirrors the IF and THEN capability we used earlier in the LAG example. The IFC and IFN functions are quite useful as they help the coding to be quite compact while delivering the required result. This helps reduce the processing time and debugging, if required. We will revisit the COALESCE function to showcase the IFN and IFC functions. The only difference between them is that the IFN function expects numeric arguments and resolves to a number, whereas the IFC function can take the first argument as numeric but expects the rest of the arguments to be of character format and resolves to a character.

Let's evaluate the differences in output using IFC and IFN:

```
Data Logical;
Input A $ B $ C $ X Y Z;
Char = COALESCEC (A, B, C);
Num = COALESCE (X, Y, Z);
IFC = IFC (A=Char, "A", "B or C");
IFN = IFN (X=Num, "X", IFN(Y=Num, "Y", "Z"));
IFN_alt = IFN (X=Num, 9, IFN(Y=Num, 99, 999));
Datalines;
FromA FromB FromC .   2 3
;
```

This will result in the following output:

Obs	A	B	C	X	Y	Z	Char	Num	IFC	IFN	IFN_alt
1	FromA	FromB	FromC	.	2	3	FromA	2	A	.	99

Using COALESCE, we tried to find the variable that matches the output. Our IFC condition specified that if the output is equal to the observation of the A variable, then output A, otherwise, output B or C. In the case of the IFN variable, we gave two IF THEN conditions. If the output for NUM is not equal to X, we checked whether the output for Y is equal to NUM. The IFC or IFN condition can take up to four arguments at any one time, but it is possible to put multiple IFC or IFN conditions in the same statement.

The only difference between the IFN and IFN_alt variables is that the arguments that are supplied for the output are characters for IFN and numeric for IFN_alt. Earlier, we discussed that all of the arguments for the IFN function need to be numeric. Hence, IFN has the output as missing whereas IFN_alt produces the desired output.

We will attempt to get similar results using a longer process to showcase the benefits of the IFC and IFN functions in the following code block:

```
Data Similar (Drop=A B C X Y Z);
Set Logical;
IF Char = "FromA" THEN LongWayC = "A";
ELSE LongWayC = "B or C";
IF Num = . THEN LongWayN = .;
ELSE IF Num = 2 THEN LongWayN = 99;
ELSE LongWayN = 999;
Run;
```

This will result in the following output:

Char	Num	IFC	IFN	IFN_alt	LongWayC	LongWayN
FromA	2	A	.	99	A	99

Similar results are produced using more statements when compared to the IFN and IFC illustration.

WHICHC or WHICHN

Earlier, we said the INDEX function searches for the first occurrence of the character string as a substring. The WHICHC and WHICHN functions search across a list of arguments and return the index of the first one that matches a given reference value. The reference value can only be specified as the first argument. There is no limit on the additional number of arguments that can be specified.

While WHICHC only works with character values and WHICHN only works with numeric values, both of them output a numeric value. If the first argument matches, it returns a value of 1; if the second argument matches, it returns a value of 2, and so on.

Let's understand the difference by looking at an example:

```
Data _NULL_;
Char = WhichC ("FromA", "FromB", "FromC", "FromA");
Char = WhichC ("FromA", "FromB", "FromC", "FromA");
Num = WhichN (100/25, 34, 4, 40, 10);
Zero = WhichC ("FromA", "FromB", "FromC", "From A");
Put Char= / Num= / Zero=;
Run;
```

This will result in the following message written to LOG:

```
Char=3
Num=2
Zero=0
NOTE: DATA statement used (Total process time):
      real time             0.00 seconds
      cpu time              0.01 seconds
```

In the case of the CHAR variable, the first argument matches the fourth argument. As the search starts from the second argument, the value that's returned is 3. The first argument is the reference value and hence the counting of the value that's returned starts only from the second argument. The NUM variable returns a value 2 as 100/25 resolves to 4 and matches the third argument. In the case of ZERO, there is no match to the reference value as the fourth argument looks similar but has a minor difference in terms of the reference value.

Up until now, we have seen the use of the PUT statement, where we wrote multiple statements to ensure that the required values are written to LOG. However, in the preceding example, a single statement is showcased that can write multiple variable values to LOG.

CHOOSEC or CHOOSEN

CHOOSEC and CHOOSEN are functions that help us to select a single value from multiple observations. The CHOOSEC (for character variables) and CHOOSEN (for numeric variables) functions come in handy when you want to use macros.

The first argument is the index value. There is no limit to the arguments supplied. For instance, in the following examples, we used up to six arguments:

SAS Statement	Results
Choosen (5, 1, 4, 5, 6, 8);	8
Choosec (-1, "A", "B", "C");	C
Choosec (3, "A", "B", "C");	C

Unlike the WHICHC or WHICHN functions, the result that's produced by CHOOSEC and CHOOSEN is not always numeric. It is numeric for CHOOSEN and a character type for CHOOSEC. However, the index value is always numeric for both CHOOSEN and CHOOSEC. In the first instance that we used CHOOSEN, please don't expect that, since the index value is 5, the result will be 5. The index value of 5 means that the fifth argument will be the output. In the case that the index value is negative, the search for the argument that's needed starts from the right and moves to the left. Hence, the C value is returned. In the last example, we require the third value and hence again we get the C value, even though scanning for the argument happens from left to right.

Number manipulation

Various databases store numbers differently. While producing reports, the format of the number might need to be changed.

CEIL, FLOOR, INT, and ROUND

While we have focused a lot on string-based functions in this chapter, there are also multiple instances when some form of treatment needs to be applied to numeric variables. The troika of ceiling, flooring, and rounding functions are the basic functions that come in handy. While some experienced SAS users may find these functions to be too basic to justify spending time practicing them, by the end of this section, you will have discovered some common mistakes that can only be avoided by the proper use of these functions.

We will use all four functions in the following code:

```
Data Decimal;
Input Score;
Ceil = Ceil (Score);
Floor = Floor (Score);
Int = Int (Score);
Round = Round (Score);
Datalines
67.454
53.34
45.23
60.80
80.4
68.5
68.9
;
```

This will result in the following output:

Obs	Score	Ceil	Floor	Int	Round
1	67.454	68	67	67	67
2	53.340	54	53	53	53
3	45.230	46	45	45	45
4	60.800	61	60	60	61
5	80.400	81	80	80	80
6	68.500	69	68	68	69
7	68.900	69	68	68	69

The output for the ceiling and floor is quite straightforward. The CEIL function resolves as the next largest integer compared to the given number. The FLOOR function works in the opposite way and converts the given number into the next smallest integer. The INT function returns the integer portion of the argument if the value of the argument is within *1E-12* of an integer. For a positive value, INT will always give the same answer as the FLOOR function. The Round function, on the other hand, rounds to the nearest integer.

For the 4, 6, and 7 observations, INT and Round produce different results. Regarding the INT function, it does not matter whether the number is closer to the next or the previous integer. It will always return the integer value given the condition we mentioned earlier. However, for the Round function, it does matter if the number is closer to the previous integer or next integer or in the middle of two integers.

We will now look at the negative numbers and observe the output:

```
Data Negative (Drop = Score);
Set Decimal;
ScoreNeg = Score*-1;
Ceil = Ceil (ScoreNeg);
Floor = Floor (ScoreNeg);
Int = Int (ScoreNeg);
Round = Round (ScoreNeg);
Run;
```

This will result in the following output:

Obs	Ceil	Floor	Int	Round	ScoreNeg
1	-67	-68	-67	-67	-67.454
2	-53	-54	-53	-53	-53.340
3	-45	-46	-45	-45	-45.230
4	-60	-61	-60	-61	-60.800
5	-80	-81	-80	-80	-80.400
6	-68	-69	-68	-69	-68.500
7	-68	-69	-68	-69	-68.900

Notice the rounding effect on the last two observations. The result is a smaller number. The Round function resolves 68.5 and -68.5 as 69 and -69 respectively. The sign has no bearing on the resolution. This may be a bit of a surprise to you as you may have expected the resolution of -68.5 as -68, that is, a move toward an increase owing to the rounding off effect.

Apart from the preceding misconception, a bigger misconception about the Round function is that the output is always an integer. The Round function allows multiple arguments that can help us round to a particular value.

Let's understand the use of Round by creating some new variables:

```
Data Fource_Round;
Thousand = Round (1564.46, 1000);
Hundreds = Round (1564.46, 100);
Tens = Round (1564.46, 10);
Unit = Round (1564.46, 1);
Tenth = Round (1564.46, .1);
Hundredth = Round (1564.46, .01);
Run;
```

This will result in the following output:

Obs	Thousand	Hundreds	Tens	Unit	Tenth	Hundredth
1	2000	1600	1560	1564	1564.5	1564.46

In this instance, the ROUND function helps us find the nearest Thousand, Hundreds, Tens, Unit, Tenth, and Hundredth number. The option to round to a specific number can be helpful when you have a bunch of numbers and you want to know how far they are from a benchmark number.

Summary

In this chapter, we presented some data problems and solved them using SAS functions. Some new topics such as LOOPS and BY GROUP processing were introduced to support the introduction of various functions. The functions that were dealt with in this chapter are primarily related to data quality and transformation. These are some of the basic tasks that an individual working with data is tasked to do. We also showcased how the same task can be completed using multiple functions. This was done to highlight how some functions can help streamline the process of writing efficient code.

In the next chapter, we will learn how to combine datasets and index, encrypt, and compress them.

Section 2: Merging, Optimizing, and Descriptive Statistics

This part discusses various methods for merging datasets, along with how to handle big data using indexing and compression techniques. Readers are introduced to the encryption functionality to safeguard their data. Descriptive statistics are generated using in-built functions.

This section comprises the following chapters:

- Chapter 3, *Combining, Indexing, Encryption, and Compression Techniques Simplified*
- Chapter 4, *Power of Statistics, Reporting, Transforming Procedures, and Functions*

3
Combining, Indexing, Encryption, and Compression Techniques Simplified

So far, we have focused on creating datasets and using multiple functions. With a plethora of datasets, the one thing that you want to do at times is combine them. A typical database will have thousands of datasets. Some larger organizations already have millions of datasets. But why would we need to combine them? If you are working for the tax authorities, you might want to merge the information you hold on individuals with the current account information you may have received from various banks. We will explore various combining techniques in this chapter.

Combining large datasets requires a lot of computing resources and takes longer to complete. This is where indexing comes in handy. We have already looked at the compress function, but there is another compression technique for datasets that helps increase efficiency when it comes to storing customer datasets. The encryption of datasets is another issue that is plaguing the data management industry. We will also look at indexing, compression, and encryption in detail within this chapter.

We will cover the following topics in this chapter:

- Introduction to combining
- Concatenating datasets
- Interleaving datasets
- Merging datasets
- Creating an index
- Encrypting datasets

Introduction to combining

The five types of combining options that are available when using a data step are as follows:

- Concatenating
- Interleaving
- Merging:
    - One-to-one
    - One-to-many
    - Many-to-many
- Updating
- Modifying

Let's go over how the datasets would be combined using the various options.

Concatenating

We will use data similar to the Cost of Living dataset that we used in `Chapter 1`, *Introduction to SAS Programming*.

The following are the two datasets:

City	Index	City	Index
Adelaide	85	Hong Kong	83
Beijing	90	Johannesburg	35
Copenhagen	65	Manila	41
Doha	56	Moscow	48
Dubai	75	Mumbai	83
Dublin	45	Munich	65

Use the following code to concatenate the two tables:

```
Data Concatenate_AB;
      Set A B;
Run;
```

This will give us the following output:

City	Index
Adelaide	85
Beijing	90
Copenhagen	65
Doha	56
Dubai	75
Dublin	45
Hong Kong	83
Johannesburg	35
Manila	41
Moscow	48
Mumbai	83
Munich	65

While concatenating, we try to stack observations of different datasets. There are complexities around the length of the observations, the different number of variables in datasets, and missing values that alter the resolution of concatenation. We will study these in detail a bit later.

Interleaving

In interleaving, we stack up datasets in a particular order. In this instance, we have specified the order through the `By Index` statement:

Index	Index
45	35
56	41
65	48
75	65
85	83
90	83

Use the following code to interleave the two tables:

```
Data Interleave_AB;
Set A B;
By Index;
Run;
```

This will result in the following output:

Index
35
41
45
48
56
65
65
75
83
83
85
90

Whenever you specify a By statement, you need to ensure that the variable that's mentioned in the statement is sorted. You may have noticed that duplicate values have been retained in the output.

Merging

Here, we will look at an instance of one-to-one merging and explore the other merges later in this chapter. First, we will run the following code to change our dataset to ensure we have two different columns, but can still leverage the data we have used for illustration purposes:

```
Data X   (Rename=(Index=IndexA));
Set A;
Run;
```

Now, we will run the `Merge` code:

```
Data Merge_XB;
Merge X B;
Run;
```

This will give us the following resultant output:

IndexA	Index
45	35
56	41
65	48
75	65
85	83
90	83

When compared to concatenation and interleaving, the key syntax difference is that you use the `Merge` statement. In terms of the output, the key difference from the preceding techniques we witnessed is that the merge statement has combined observations based on their position in the datasets. If the variable name was kept the same in both datasets, the output would only contain the observations from the second or the last dataset that was read in the merge statement.

Updating

We will reuse the Table A data that was used for concatenation from the concatenation section here. The `A_Alt` table will be used to update the original dataset, that is, A.

Use the following code to update Table A:

```
Data A_Alt;
Input City $12. Index;
Datalines;
Adelaide      85
Beijing       90
Copenhagen    .
Copenhagen    65
Dubai         75
Dublin        95
Hong Kong     83
;
```

```
Data A;
Update A A_Alt;
By City;
Run;
```

This will result in the following output:

City	Index	City	Index	City	Index
Adelaide	85	Adelaide	85	Adelaide	85
Beijing	90	Beijing	90	Beijing	90
Copenhagen	65	Copenhagen	.	Copenhagen	65
Doha	56	Copenhagen	65	Doha	56
Dubai	75	Dubai	75	Dubai	75
Dublin	45	Dublin	95	Dublin	95
		Hong Kong	83	Hong Kong	83

Here, we attempted to update Table A with A_Alt. Observe the following:

- The number of observations is different
- Observation 1 and 2 in A_Alt have different values for Adelaide and Beijing
- Observation 3 in A_Alt has a missing value
- Observation 3 and 4 in A_Alt both relate to the same city
- Doha is not listed in A_Alt but is present in Table A
- Hong Kong is not listed in A but is present in Table A_Alt

This disparity has been resolved like so in the output:

- Values for observation 1 and 2 have been updated
- The missing observation, 3, in A_Alt for Copenhagen has been disregarded
- It has retained only one value for Copenhagen
- It has retained Doha but also added Hong Kong to the list of cities

Modifying

There are a couple of key differences between modifying and updating. In modifying, we can only modify the existing table; we cannot create a new output table. For modifying, we do not need the observations to be sorted, whereas it is mandatory for updating.

The following code is used to modify the table:

```
Data Master;
Modify Master A_Alt;
By City;
Run;
```

This will result in the following output:

Obs	City	Index		Obs	City	Index		Obs	City	Index
1	Adelaide	85		1	Adelaide	65		1	Adelaide	65
2	Beijing	90		2	Beijing	98		2	Beijing	98
3	Copenhagen	65		3	Copenhagen	.		3	Copenhagen	.
4	Doha	56		4	Copenhagen	65		4	Copenhagen	65
5	Dubai	75		5	Dubai	75		5	Dubai	75
6	Dublin	45		6	Dublin	95		6	Dublin	95
				7	Hong Kong	83		7	Hong Kong	83

The modification has been requested for the Master dataset. The output dataset is a replica of the A_Alt table. If we use the **REPLACE** option in the data statement, we get a different answer, as follows:

Obs	City	Index
1	Adelaide	65
2	Beijing	98
3	Copenhagen	65
4	Copenhagen	65
5	Dubai	75
6	Dublin	95
7	Hong Kong	83

The **REPLACE** statement writes the current observation to the same physical location from which it was read in all the datasets that are named in the data statement.

Now that we've introduced a plethora of options that we can use to combine datasets, it's important that we now delve deeper into them.

Concatenation

When we were introduced to concatenation, we looked at the simplest of datasets. Both datasets had variables with similar attributes, equal variable length, and the same type of variables. We ended up with a result that had the two datasets stacked together.

Different variable length and additional variables

Now, we will look at some small dataset differences that we need to overcome in the practical world. As part of the deep dive, we will also look at the APPEND methodology to concatenate datasets. Let's get started:

```
Data Customer;
Set Customer_X Customer_Y;
Run;
```

This will give us the following tables as output:

Obs	ID	Gender	Age	Region
1	10004523	F	34	Portsmouth
2	10002342	F	45	Southampton
3	10002462	M	36	Leeds
4	10002328	M	65	Durham
5	10006345	M	56	Bristol
6	10005234	M	19	Newcastle
7	10005325	F	23	London

Obs	ID	Gender	Age	Region	Dependents
1	10005296	F	24	Shefield	1
2	10001002	F	65	Liverpool	0
3	10003407	F	43	Cardiff	0
4	10009832	M	76	Bath	0
5	10000086	F	21	Sunderland	0
6	10002349	M	27	London	2
7	10008740	M	40	Birmingham	3

Note that we have missing values for the **Dependents** variable in the output:

Obs	ID	Gender	Age	Region	Dependents
1	10004523	F	34	Portsmouth	.
2	10002342	F	45	Southampton	.
3	10002462	M	36	Leeds	.
4	10002328	M	65	Durham	.
5	10006345	M	56	Bristol	.
6	10005234	M	19	Newcastle	.
7	10005325	F	23	London	.
8	10005296	F	24	Shefield	1
9	10001002	F	65	Liverpool	0
10	10003407	F	43	Cardiff	0
11	10009832	M	76	Bath	0
12	10000086	F	21	Sunderland	0
13	10002349	M	27	London	2
14	10008740	M	40	Birmingham	3

A quick look at the LOG shows the following comment:

```
WARNING: Multiple lengths were specified for the variable Region by input
data set(s). This can cause truncation of data.
 NOTE: There were 7 observations read from the data set WORK.CUSTOMER_X.
 NOTE: There were 7 observations read from the data set WORK.CUSTOMER_Y.
 NOTE: The data set WORK.CUSTOMER has 14 observations and 5 variables.
```

We got a note about the length because, in Customer_X, the length of Region has been defined as 11 characters, whereas it has been defined as 12 in Customer_Y. We would not have got an issue if the length in Customer_Y was less than it was in the Customer_X table. The only reason it didn't actually cause truncation in the output dataset is that the maximum length of the observations in the Customer_Y table is 10.

There was, however, no WARNING or NOTE regarding the additional variable in the Customer_Y dataset. The additional variable is written to the output dataset but the observations from Customer_X for the additional variable are set to missing when the two tables are stacked together.

So, what should you do if you get a note about differing lengths of variables in datasets? Two solutions will be offered to you but a dilemma will also be posed for one of the solutions:

Obs	ID	Region		Obs	ID	Region		Obs	ID	Region
								1	10004523	Bath
								2	10002342	Leed
1	10004523	Bath		1	100296	Newcastle		3	100296	Newca
2	10002342	Leed		2	101002	Birmingham		4	101002	Birmi

In Table X, **ID** has a length of 8 and **Region** a length of 5. In Table Y, **ID** has a length of 6 and **Region** a length of 10. If the issue for the difference in length was for the **ID** variable only, we could have ignored the disparity between the X and Y datasets and could be content with the output in the XY dataset. After all, no truncation of values for the ID variables in the Y dataset would have occurred as the variable has a shorter length and shorter observations. However, considering that the observations of the **Region** variable in the Y dataset have been truncated, we would need to correct this situation.

One simple solution to prevent the truncation of the Region variable would be to change the order of the datasets in the SET statement. Use the following concatenation command:

```
Data XY;
Set Y X;
Run;
```

This will give the following table as the resultant output:

Obs	ID	Region
1	100296	Newcastle
2	101002	Birmingham
3	10004523	Bath
4	10002342	Leed

The issue of truncation has been dealt with for the **Region** variable. If you look at the properties/attributes of the **ID** variable, you will notice that the length of the variable is 6. Even though the **ID** values of Y are displayed properly, the length of the observations from the table has changed from 8 to 6 after concatenation. Let's attempt to change the length values using the following commands:

```
Data XY;
Set Y X;
Length ID 8.;
Length Region $12.;
Run;

WARNING: Length of character variable Region has already been set.
Use the LENGTH statement as the very first statement in the DATA STEP to
declare the length of a character variable.
```

The preceding code generates the following output:

Property	Value	Property	Value
Label	ID	Label	Region
Name	ID	Name	Region
Length	8	Length	10
Type	Numeric	Type	Char
Format		Format	

The preceding screenshot contains the program, LOG, and the output from concatenation. As you can see, the ID variable's length has been changed successfully. For the character variable, we got a warning stating the length has already been specified following the invocation of the **SET** statement. If we want to change the length of the character variable, then we should do it before invoking the **SET** statement. For a numeric value, it doesn't matter whether the length is changed prior to or post the invocation of the **SET** statement. Programmers need to be careful as making the mistake of placing the **LENGTH** statement in an incorrect order can lead to unintended consequences. It is not uncommon for programmers to miss the **LOG** message as the output gets created regardless.

Duplicate values

What if some of the observations in tables the same values? In that case, the output dataset would have duplicate values. The concatenation of the does not lead to the removal of any values as long as the output dataset is created successfully.

Different data types

The ID variable in both datasets we concatenated has a numeric format. Let's see what happens when it is declared as a character variable in the Customer_Y dataset:

```
103 Data Customer;
 104 Set Customer_X Customer_Y;
 ERROR: Variable ID has been defined as both character and numeric.
 105 Run;

 NOTE: The SAS System stopped processing this step because of errors.
 WARNING: The data set WORK.CUSTOMER may be incomplete. When this step was
 stopped there were 0 observations and 5 variables.
```

The only way out for us is to change the format of the ID in either of the datasets. The decision on which dataset needs to have the format changed should depend on what values are expected for the ID variable. If only numeric values are expected, then the format of the ID variable in the Customer_Y dataset should be changed, otherwise, the ID variable format in the Customer_X dataset should be changed.

Leveraging the temporary variable

Take another look at the table in the *Different variable length and additional variables* section. There are five variables in the output dataset. It would be helpful if we knew which observations came from the Customer_X or Customer_Y table. We will modify the program to create a Source variable that tells us which table has contributed to specific observations:

```
Data Customer;
Set Customer_X (in = a) Customer_Y (in = b);
If a = 1 then
    Source = "X";
Else Source = "Y";
Run;
```

This will give the following table as the resultant output:

Obs	ID	Gender	Age	Region	Dependents	Source
1	10004523	F	34	Portsmouth	.	X
2	10002342	F	45	Southampton	.	X
3	10002462	M	36	Leeds	.	X
4	10002328	M	65	Durham	.	X
5	10006345	M	56	Bristol	.	X
6	10005234	M	19	Newcastle	.	X
7	10005325	F	23	London	.	X
8	10005296	F	24	Shefield	1	Y
9	10001002	F	65	Liverpool	0	Y
10	10003407	F	43	Cardiff	0	Y
11	10009832	M	76	Bath	0	Y
12	10000086	F	21	Sunderland	0	Y
13	10002349	M	27	London	2	Y
14	10008740	M	40	Birmingham	3	Y

The `IN` temporary variable is only available for the duration of the data step. However, by using it, we can create a permanent variable, such as Source. The form of the `IN` temporary variable is as follows:

$$IN = varname$$

The variable can only take values of 1 or 0. In the preceding instance, all the variables that are from `Customer_X` will take the value of 1 for the A variable. In the preceding program, we have the following:

```
If a = 1 then
    Source = "X";
```

We can write the following instead:

```
If a then
    Source = "X";
```

Having a source variable is helpful in concatenation as identifying the source gets confusing even with two contributing tables. Imagine the difficulty if 10 tables had been used in the concatenation process.

PROC APPEND

PROC APPEND is an alternate way of stacking up datasets. It differs from concatenation since it uses the SET statement in the following ways:

- PROC APPEND can only handle two datasets, whereas the only limitation for concatenation is the computing resources.
- In PROC APPEND, you cannot create a new dataset.
- PROC APPEND is the most helpful for large datasets as it does not read the dataset and merely appends the second dataset to the first.
- The PROC APPEND procedure can be forced to create an output when the variables in datasets are of different types. Concatenation via the SET statement will stop processing and not produce an output.

We will reuse the dataset from the *Concatenating* section and get familiar with the syntax of PROC APPEND:

```
PROC APPEND Base = A Data = B;
RUN;
```

The following table is the resultant output:

Obs	City	Index
1	Adelaide	85
2	Beijing	90
3	Copenhagen	65
4	Doha	56
5	Dubai	75
6	Dublin	45
7	Hong Kong	83
8	Johannesburg	35
9	Manila	41
10	Moscow	48
11	Mumbai	83
12	Munich	65

We have defined the data that will be appended to using the BASE syntax. The result that was produced was the same as we got for concatenating.

Let's check the output when we have different variable types. We tried to use the datasets we showcased in the different variable length. No output was created and we got the following message in the LOG:

```
103 PROC APPEND Base = Customer_X Data = Customer_Y;
 104 RUN;

NOTE: Appending WORK.CUSTOMER_Y to WORK.CUSTOMER_X.
 WARNING: Variable Dependents was not found on BASE file. The variable will
not be added to the BASE file.
 WARNING: Variable ID not appended because of type mismatch.
 WARNING: Variable Region has different lengths on BASE and DATA files
(BASE 11 DATA 12).
 ERROR: No appending done because of anomalies listed above. Use FORCE
option to append these files.
 NOTE: 0 observations added.
 NOTE: The data set WORK.CUSTOMER_X has 7 observations and 4 variables.
```

As per the suggestion, we are going to use the FORCE option:

```
PROC APPEND Base = Customer_X Data = Customer_Y FORCE;
RUN;
```

The following table is the resultant output:

Obs	ID	Gender	Age	Region
1	10004523	F	34	Portsmouth
2	10002342	F	45	Southampton
3	10002462	M	36	Leeds
4	10002328	M	65	Durham
5	10006345	M	56	Bristol
6	10005234	M	19	Newcastle
7	10005325	F	23	London
8	.	F	24	Shefield
9	.	F	65	Liverpool
10	.	F	43	Cardiff
11	.	M	76	Bath
12	.	F	21	Sunderland
13	.	M	27	London
14	.	M	40	Birmingham

There are three differences between the `Customer_X` and `Customer_Y` tables. `Customer_X` has `ID` as numeric, whereas `Customer_Y` has it stored as a character. The length of `Region` for`Customer_X` and `Customer_Y` are 11 and 12, respectively. `Customer_Y` has the `Dependent` variable, whereas `Customer_X` doesn't.

In the case of the variable type difference and the length of the variable, the output for the ID and Region columns takes into account the attributes of the BASE dataset. The output only includes the variable in the BASE table and ignores the additional variable in the table to be appended. Even with the FORCE option, the additional variable could not be written to the BASE dataset as it is not contained in the descriptor portion of the `Customer_X` dataset. Remember that we are not creating a new dataset with PROC APPEND. We are merely appending it to the BASE dataset, and hence the descriptor portion that we studied in earlier chapters becomes important in this instance. If we had an extra variable in the BASE dataset compared to the appended dataset, with the FORCE option, we would have been able to get an output. The observations for the appended dataset would have the values set to missing for the additional variable contained in the BASE dataset.

The FORCE option ensures that the length and variable type in the BASE dataset supersedes the information in the appended dataset. If an additional variable is present in the BASE dataset, then the variable is present in the appended output.

Given all of these differences, if the dataset you're going to be appending is large and if you don't have issues with it being overwritten, then PROC APPEND might be a better option than concatenation via the SET statement.

Interleaving

To showcase how Interleave handles variations in length, data type, and additional variables, we have modified the data we used earlier for introducing interleave. The program that we used to interleave is still the same:

```
Data A;
Input Index City $1. Sample Past;
Datalines;
45 A 500 43
56 B 500 50
65 C 600 58
75 D 600 68
85 E 600 82
90 F 500 94
;
```

```
Data B;
Input Index City $2. Sample $;
Datalines;
35 AA 600
41 BB 500
48 CC 500
65 DD 600
83 EE 600
83 FF 600
;
```

Let's review the datasets that were produced:

Index	City	Sample	Past	Index	City	Sample
45	A	500	43	35	AA	600
56	B	500	50	41	BB	500
65	C	600	58	48	CC	500
75	D	600	68	65	DD	600
85	E	600	82	83	EE	600
90	F	500	94	83	FF	600

The preceding dataset's Index variable only has the same attributes. City is a character variable type in both datasets but the length is different. For Sample, the variable type is different in the datasets. The Past variable only exists in dataset A.

When we run the interleaving code, we get the following error:

```
101 Data Interleave_AB;
102 Set A B;
ERROR: Variable Sample has been defined as both character and numeric.
103 By Index;
104 Run;
```

To resolve this, we will need to create a consistent variable type across both datasets. Let's run the program again after converting Sample into numeric format.

The following table is the resultant output that you would get:

Index	City	Sample	Past
35	A	600	.
41	B	500	.
45	A	500	43
48	C	500	.
56	B	500	50
65	C	600	58
65	D	600	.
75	D	600	68
83	E	600	.
83	F	600	.
85	E	600	82
90	F	500	94

The issue with Sample is a thing of the past. The City variable has retained the length in dataset A and this has resulted in the truncation of values of the observation from dataset B. Past is a variable that is only present in the first dataset in the SET statement and hence the new dataset has been created with the Past variable. All the observations from dataset B have the value set to missing for the variable.

Merging

While an appreciation of different ways of combining datasets is necessary, the most important methodology in SAS is merging datasets. It is time to look at one-to-many and many-to-many merges. Along with these two types of merges, we will also look at the concept of BY MATCHING.

By Matching

For performing By Matching, we have the following information about the cost of living in two different datasets, A and B, at hand:

City	Index	Prev_yr_index	Housing	Food	Travel	City	Utility	Education	Leisure	Other
Adelaide	85	83	35	10	10	Adelaide	9	14	10	12
Beijing	90	92	40	10	15	Beijing	10	18	5	2
Copenhagen	65	64	25	15	10	Copenhagen	10	12	12	16
Doha	56	50	30	15	5	Doha	10	10	20	10
Dubai	75	76	30	16	14	Dubai	10	20	8	2
Dublin	45	43	30	10	8	Dublin	12	10	15	15
Hong Kong	83	88	45	5	10	Hong Kong	15	15	9	1
Johannesburg	35	40	45	5	5	Johannesburg	15	15	10	5
Manila	41	42	25	10	15	Manila	15	20	10	5
Moscow	48	53	40	20	5	Moscow	5	10	10	10

We want to join the two datasets together so that we have one wide dataset with 10 rows of observations for City and nine variables.

Let's use the same form of Merge that we used earlier when generating the output for Merging:

```
Data Cost_Living;
Merge A B;
Run;
```

We get the desired output in the following table. The two datasets have been combined based on the order of the observations:

City	Index	Prev_yr_index	Housing	Food	Travel	Utility	Education	Leisure	Other
Adelaide	85	83	35	10	10	9	14	10	12
Beijing	90	92	40	10	15	10	18	5	2
Copenhagen	65	64	25	15	10	10	12	12	16
Doha	56	50	30	15	5	10	10	20	10
Dubai	75	76	30	16	14	10	20	8	2
Dublin	45	43	30	10	8	12	10	15	15
Hong Kong	83	88	45	5	10	15	15	9	1
Johannesburg	35	40	45	5	5	15	15	10	5
Manila	41	42	25	10	15	15	20	10	5
Moscow	48	53	40	20	5	5	10	10	10

But what if the number of observations wasn't the same in both datasets? Let's remove the observations of Doha and Dubai in dataset B and try to rerun our program. We now get the following output:

Obs	City	Index	Prev_yr_index	Housing	Food	Travel	Utility	Education	Leisure	Other
1	Adelaide	85	83	35	10	10	9	14	10	12
2	Beijing	90	92	40	10	15	10	18	5	2
3	Copenhagen	65	64	25	15	10	10	12	12	16
4	Dublin	56	50	30	15	5	12	10	15	15
5	Hong Kong	75	76	30	16	14	15	15	9	1
6	Johannesburg	45	43	30	10	8	15	15	10	5
7	Manila	83	88	45	5	10	15	20	10	5
8	Moscow	35	40	45	5	5	5	10	10	10
9	Manila	41	42	25	10	15	.	.	.	.
10	Moscow	48	53	40	20	5	.	.	.	.

Instead of putting missing values for **Doha** and **Dubai**, the two city names have been omitted from the **City** variable. Furthermore, there is a duplication of names for **Manila** and **Moscow**. **Manila** and **Moscow** have an incorrect set of values for all the variables for observations **7** and **8**.

To solve this, we will have to use By Matching. We used the By keyword in the interleave examples earlier. Merging with a By statement allows the observations to be merged according to the values of the By variables specified. All the datasets involved must be sorted by the variables that are going to be specified in the By statement:

```
Data Cost_Living;
Merge A B;
By City;
Run;
```

The preceding code will give us the following table as the resultant output, which is the cost of living dataset that was obtained using By Match merging:

Obs	City	Index	Prev_yr_index	Housing	Food	Travel	Utility	Education	Leisure	Other
1	Adelaide	85	83	35	10	10	9	14	10	12
2	Beijing	90	92	40	10	15	10	18	5	2
3	Copenhagen	65	64	25	15	10	10	12	12	16
4	Doha	56	50	30	15	5	.	.	.	.
5	Dubai	75	76	30	16	14	.	.	.	.
6	Dublin	45	43	30	10	8	12	10	15	15
7	Hong Kong	83	88	45	5	10	15	15	9	1
8	Johannesburg	35	40	45	5	5	15	15	10	5
9	Manila	41	42	25	10	15	15	20	10	5
10	Moscow	48	53	40	20	5	5	10	10	10

We get the correct answer this time around, with the **Utility**, **Education**, **Leisure**, and **Other** variables getting their values set to missing for **Doha** and **Dubai**.

Overlapping variables

While introducing one-to-one merging earlier in this chapter, we mentioned that if the variable name and its attributes were the same across two datasets that were merged, the output would only contain the observations from the second or the last dataset that was read in the merge statement. Let's look at an example to understand its implications.

We will still use the same Merge statement we used in the previous example. Our dataset, B, as shown in the following table, is marginally different from the previous example as it now contains an additional variable, **Travel**:

Obs	City	Utility	Education	Leisure	Other	Travel
1	Adelaide	9	14	10	12	.
2	Beijing	10	18	5	2	.
3	Copenhagen	10	12	12	16	99
4	Dublin	12	10	15	15	99
5	Hong Kong	15	15	9	1	8
6	Johannesburg	15	15	10	5	7
7	Manila	15	20	10	5	5
8	Moscow	5	10	10	10	8

The output that was generated in the following table by the merging process does not create errors or warnings in the log. All the observations of the Travel dataset, apart from Doha and Dubai, have been overwritten by the values of the same variable in dataset B. The values of Travel for observation 1, 2, 3, and 6 are of particular interest as they are either missing or seemingly erroneous:

Obs	City	Index	Prev_yr_index	Housing	Food	Travel	Utility	Education	Leisure	Other
1	Adelaide	85	83	35	10	.	9	14	10	12
2	Beijing	90	92	40	10	.	10	18	5	2
3	Copenhagen	65	64	25	15	99	10	12	12	16
4	Doha	56	50	30	15	5	.	.	.	.
5	Dubai	75	76	30	16	14	.	.	.	.
6	Dublin	45	43	30	10	99	12	10	15	15
7	Hong Kong	83	88	45	5	8	15	15	9	1
8	Johannesburg	35	40	45	5	7	15	15	10	5
9	Manila	41	42	25	10	5	15	20	10	5
10	Moscow	48	53	40	20	8	5	10	10	10

One of the issues plaguing most data systems in large organizations is the presence of similar variables across datasets. It is a fallacy to assume that the values will be similar across the datasets. These values could be from a different time period or currencies, the sources may be different, or they may be transformed after performing some data quality checks. The reasons for two similar-looking variables across datasets can be endless. The analyst performing the Merge should be careful about the consequences of using overlapping variables. Since no warning or errors are produced on executing the program, it is easy to combine data incorrectly.

One-to-many merging

Compared to the one-to-one matching via BY Matching, there is no change in syntax required for one-to-many merging. The difference is in the input and output datasets. The following tables are the datasets that we will merge in one-to-many merging:

Obs	Index_Date	City	Index
1	01JAN2019	Adelaide	85
2	01JAN2019	Beijing	90
3	01JAN2018	Beijing	89
4	01JAN2019	Copenhagen	65
5	01JAN2019	Dublin	45
6	01JAN2019	Hong Kong	83
7	01JAN2018	Hong Kong	81
8	01JAN2017	Hong Kong	76

Obs	City	Index	Housing	Food
1	Adelaide	85	83	35
2	Beijing	90	92	40
3	Copenhagen	65	64	25
4	Dublin	45	43	30
5	Hong Kong	83	88	45

Obs	City	Utility	Education
1	Adelaide	9	14
2	Beijing	10	18
3	Copenhagen	10	12
4	Dublin	12	10
5	Hong Kong	15	15

We have introduced a third dataset here, C, which contains multiple observations for some cities. It contains the index values of previous years as well. Up until now, we had a separate variable (instead of a row) that contained the previous index values in the various examples in this chapter. We have a one row per city situation in datasets A and B, but a multiple rows per city situation in dataset C. Let's look at the outcome of the Merge process:

```
Data ABC;
Merge A B C;
By City;
Run;
```

The preceding code will give the following table as the resultant output:

Obs	City	Index	Housing	Food	Utility	Education	Index_Date
1	Adelaide	85	83	35	9	14	01JAN2019
2	Beijing	90	92	40	10	18	01JAN2019
3	Beijing	89	92	40	10	18	01JAN2018
4	Copenhagen	65	64	25	10	12	01JAN2019
5	Dublin	45	43	30	12	10	01JAN2019
6	Hong Kong	83	88	45	15	15	01JAN2019
7	Hong Kong	81	88	45	15	15	01JAN2018
8	Hong Kong	76	88	45	15	15	01JAN2017

The first instance of multiple rows for a city in dataset C is for Beijing. As a consequence, we get two observations for the city in the output dataset. The index values and dates are different for both observations. However, the rest of the variables for Beijing have the same values for both observations. This seems to be the correct solution, as storing them as missing for one of the observations is factually incorrect. Remember, for Beijing, we only had one row of observations for the Housing, Food, Utility, and Education variables in datasets A and B. The second observation for Beijing in the output contains the same values for these observations as in the first observation. Hence, in the preceding screenshot, we can see two key aspects of overlapping variables. First, the observations in the last dataset that were specified in the SET statement override any previous observations for overlapping variables. Second, the values for some variables will be repeated across rows/observations in a one-to-many match scenario.

Program data vector

We explored the concept of the **Program Data Vector** (PDV) in previous chapters. In this section, we will look at what the PDV of the previously merged dataset would look like. This will provide us with a further understanding of how the Merge process works. We will look at the roles of _N_ and _ERROR_. The focus is on understanding how observations from the three datasets will consolidate into one dataset. The following are the steps that PDVs follow:

1. Initially, the PDV will be created by incorporating the variables from datasets A, B, and C and by setting all the records to missing:

City	Index	Housing	Food	Utility	Education	Index_Date
	.	.	.	.	.	.

2. The program will look for the first BY group that should go into the PDV. In our case, the BY group is formed by the Index variable. The first BY group is for Adelaide. There are observations for this BY group in all three datasets. The observations from dataset A will be written to the PDV first:

City	Index	Housing	Food	Utility	Education	Index_Date
Adelaide	85	83	35	.	.	.

3. After this, the observations from dataset B for the BY group for Adelaide will be written to the PDV:

City	Index	Housing	Food	Utility	Education	Index_Date
Adelaide	85	83	35	9	14	.

4. It is now time for the observations from dataset C for the BY group for Adelaide to be written to the PDV. In dataset C, we have values for two variables for this BY group. Index_Date values are missing for the PDV, so they will be copied straight away. The value of Index is already there in the PDV. This will get overwritten by the value of Index in dataset C. However, since in datasets A and C the value of Index is same for the BY group Adelaide, we won't notice the difference in actual values (even though the 85 value now comes from dataset C):

City	Index	Housing	Food	Utility	Education	Index_Date
Adelaide	85	83	35	9	14	01Jan19

5. After writing this observation in the PDV, the program will check if there are any more observations for this BY group. Having found none, it will set the values in the PDV to missing after writing the PDV data into the output dataset:

City	Index	Housing	Food	Utility	Education	Index_Date
	.	.	.	.	.	.

6. The program will now look for the next BY group. The next BY group is Beijing and it has values in all three datasets. The PDV will be populated by the values in dataset A first:

City	Index	Housing	Food	Utility	Education	Index_Date
Beijing	90	92	40	.	.	.

7. At this stage, the PDV will be populated by the first observation for the Beijing BY group from dataset B:

City	Index	Housing	Food	Utility	Education	Index_Date
Beijing	90	92	40	10	18	.

8. Just like in the case of the BY group for Adelaide, we overwrite the value of Index in the PDV with the value of the variable from dataset C. However, since the values in dataset A and C for the first instance of the BY group are the same, we won't notice any difference. The value of Index_Date will also be populated:

City	Index	Housing	Food	Utility	Education	Index_Date
Beijing	90	92	40	10	18	01Jan2019

9. In the instance of the BY group for `Adelaide`, after writing the first observation in the PDV, the PDV was set to missing after the data was written into the data table. This happened because the program couldn't find any more observations for the BY group. However, in the case of the BY group for Beijing, we have two observations in dataset C. Hence, the PDV will not set the variable values to missing just yet. Instead, it will create a second observation where the values of the variable are initially set to missing:

City	Index	Housing	Food	Utility	Education	Index_Date
Beijing	90	92	40	10	18	01Jan2019
	.	.	.	.	.	.

10. Since the BY group for `Beijing` doesn't have any observations in datasets A and B, the values from the variables from the observation that was retained in the PDV will be written to the second observation:

City	Index	Housing	Food	Utility	Education	Index_Date
Beijing	90	92	40	10	18	01Jan2019
Beijing	90	92	40	10	18	.

11. Dataset C does contain a second observation for the BY group for Beijing. Its Index value will now overwrite the existing value for the variable in the second observation of the PDV. The value of `Index_Date` will also be written to the PDV:

City	Index	Housing	Food	Utility	Education	Index_Date
Beijing	90	92	40	10	18	01Jan2019
Beijing	89	92	40	10	18	01Jan2018

12. At this stage, the program will realize that there are no more observations for the current form of `Beijing`. The program will write the PDV values to the dataset, set the PDV values to missing, and continue scanning the rest of the observations for the next BY group.

Many-to-many merging

In the one-to-many merge illustration, we had multiple observations for the same BY group in dataset C. In a many-to-many merge situation, we can end up with multiple observations for the same BY group in different datasets. Let's use the same program we used in the one-to-many merging but with slightly modified datasets:

Obs	Index_Date	City	Index	Housing
1	01JAN2019	Adelaide	85	83
2	01JAN2019	Beijing	90	92
3	01JAN2018	Beijing	90	90
4	01JAN2019	Copenhagen	65	64
5	01JAN2019	Dublin	45	43
6	01JAN2019	Hong Kong	83	88
7	01JAN2018	Hong Kong	83	88
8	01JAN2017	Hong Kong	82	88
9	01JAN2016	Hong Kong	82	87

Obs	City	Utility	Education
1	Adelaide	9	14
2	Beijing	10	18
3	Copenhagen	10	12
4	Dublin	12	10
5	Hong Kong	15	15

Obs	Index_Date	City	Index	Food
1	01JAN2019	Adelaide	85	35
2	01JAN2019	Beijing	90	45
3	01JAN2018	Beijing	89	42
4	01JAN2019	Copenhagen	65	30
5	01JAN2019	Dublin	45	34
6	01JAN2019	Hong Kong	83	40
7	01JAN2018	Hong Kong	81	39
8	01JAN2017	Hong Kong	76	36

The merged output is as follows:

Obs	Index_Date	City	Index	Housing	Utility	Education	Food
1	01JAN2019	Adelaide	85	83	9	14	35
2	01JAN2019	Beijing	90	92	10	18	45
3	01JAN2018	Beijing	89	90	10	18	42
4	01JAN2019	Copenhagen	65	64	10	12	30
5	01JAN2019	Dublin	45	43	12	10	34
6	01JAN2019	Hong Kong	83	88	15	15	40
7	01JAN2018	Hong Kong	81	88	15	15	39
8	01JAN2017	Hong Kong	76	88	15	15	36
9	01JAN2016	Hong Kong	82	87	15	15	36

The merge was performed successfully as per the LOG. However, the output is incorrect in many ways. The following are a few lines of the LOG:

```
NOTE: MERGE statement has more than one data set with repeats of BY values.
NOTE: There were 9 observations read from the data set WORK.A.
NOTE: There were 5 observations read from the data set WORK.B
NOTE: There were 8 observations read from the data set WORK.C.
NOTE: The data set WORK.ABC has 9 observations and 7 variables.
 NOTE: DATA statement used (Total process time):
       real time 0.00 seconds
       cpu time 0.00 seconds
```

We will only look at observation 9 in the output. For 2017-19 observations, the values of Index are the same as the values of the variable in dataset C. However, for 2016, since there is no Index value in dataset C, the value in the output dataset has been sourced by the PDV from dataset A. Moreover, Food has been arbitrarily given the value of 36 for 2016. However, we can't say that this is true. The PDV would have retained the value of 36 before processing the fourth observation in the BY group for Hong Kong. This value got written to the dataset after we processed the fourth observation in the PDV. The LOG also didn't produce any WARNINGS or ERRORS. This, however, doesn't mean that we got the desired result.

To get the right answer, we should have done a multiple BY group match. In our case, the second BY group is `Index_Date`. Let's sort the datasets by both BY groups, as this is a necessary condition of merging, and then try to remerge the datasets:

```
Proc Sort Data = A;
By City Index_Date;
Run;

Proc Sort Data = C;
By City Index_Date;
Run;
```

We will only be using datasets A and C to match-merge on two BY groups since dataset B doesn't have one of the BY groups we have used:

```
Data AC;
Merge A C;
By City Index_Date;
Run;
```

This will give the following table as the resultant output:

Obs	Index_Date	City	Index	Housing	Food
1	01JAN2019	Adelaide	85	83	35
2	01JAN2018	Beijing	89	90	42
3	01JAN2019	Beijing	90	92	45
4	01JAN2019	Copenhagen	65	64	30
5	01JAN2019	Dublin	45	43	34
6	01JAN2016	Hong Kong	82	87	.
7	01JAN2017	Hong Kong	76	88	36
8	01JAN2018	Hong Kong	81	88	39
9	01JAN2019	Hong Kong	83	88	40

Now, we get the correct result for the 2016 record for the Food variable. The value has been set to missing since there is no observation for the year in dataset C. Dataset A doesn't contain the Food variable. If it is, its value from the previous observations of the BY group in the PDV would have got carried over to the dataset.

Did you know that match merging is case-sensitive? It can throw up unintended consequences by using it. Imagine a dataset with millions of rows of customer records derived from different data systems. All the records are sorted using the BY groups to merge. However, 100,000 records out of the millions of records have customer names in lowercase. You should be able to match this dataset to any other dataset as long as the syntax is correct and other match-merge conditions have been met. However, it's likely you won't get the desired result. Your other dataset is mostly going to be uppercase, lowercase, or proper case in terms of the customer names. This is going to cause a mismatch. This has been showcased using the smaller datasets, that is, A and C, to make it easier for you to spot the issue:

Obs	Index_Date	City	Index	Housing	Food
1	01JAN2019	Adelaide	85	83	35
2	01JAN2019	Copenhagen	65	64	30
3	01JAN2016	Hong Kong	82	87	.
4	01JAN2017	Hong Kong	82	88	.
5	01JAN2018	Hong Kong	83	88	.
6	01JAN2019	Hong Kong	83	88	.
7	01JAN2017	hong Kong	76	.	36
8	01JAN2018	hong Kong	81	.	39
9	01JAN2019	hong Kong	83	.	40

Obs	Index_Date	City	Index	Housing
1	01JAN2019	Adelaide	85	83
2	01JAN2019	Copenhagen	65	64
3	01JAN2016	Hong Kong	82	87
4	01JAN2017	Hong Kong	82	88
5	01JAN2018	Hong Kong	83	88
6	01JAN2019	Hong Kong	83	88

Obs	Index_Date	City	Index	Food
1	01JAN2019	Adelaide	85	35
2	01JAN2019	Copenhagen	65	30
3	01JAN2017	hong Kong	76	36
4	01JAN2018	hong Kong	81	39
5	01JAN2019	hong Kong	83	40

The observations for Hong Kong have not matched due to the case mismatch in both datasets. The following is the LOG for the match-merge:

```
NOTE: There were 6 observations read from the data set WORK.A.
NOTE: There were 5 observations read from the data set WORK.C.
NOTE: The data set WORK.AC has 9 observations and 5 variables.
NOTE: DATA statement used (Total process time):
      real time 0.00 seconds
      cpu time 0.01 seconds
```

Even though no WARNINGS or ERROR messages were published to the LOG regarding the case mismatch issue, it can still be a good pointer to potential match-merge problems. If we expected a good match between the two datasets, the output of 9 observations when the maximum numbers of observations were 6 in the input dataset points to a match-merge-issue.

You should consider converting the cases of the BY groups prior to Merge to avoid case sensitivity issues.

Indexing

We broached the subject of creating an index file in previous chapters. In this section, we will describe how the indexes are stored and retrieved. The index file consists of entries that are organized hierarchically and connected by pointers, all of which are maintained by SAS. The lowest level in the index file hierarchy consists of entries that represent each distinct value for an indexed variable, in ascending value order. Each entry contains the following information:

- A distinct value
- One or more unique record identifiers (referred to as a RID) that identify each observation that contains the value

If we created an index for the City variable using the AC dataset in the mismatch dataset, we would have an index file with entries such as the following:

Value	RID
Adelaide	1
Copenhagen	2
Hong Kong	3, 4, 5, 6
hong Kong	7, 8, 9

Let's say we wrote the following code:

```
Data hK;
Set AC;
Where City eq "hong Kong";
Run;
```

The output would contain observations 7, 8, and 9 from the AC dataset. When an index is present, the use of a *where* clause means that SAS would calculate the median of the observations. In this case, it will see if 5 < 7 or 5 > 7. It will remove all the observations before 5 as the RID for "hong Kong" is positioned at the value 7. SAS then moves sequentially through the index entries, reading observations until it reaches the index entry for the value that is equal to or greater than 7.

To further understand sequential processing, let's consider what will be the decision tree for SAS to try toretrieve a record from the 281st RID in a 400-RID file. The following diagram shows the route to finding the RID:

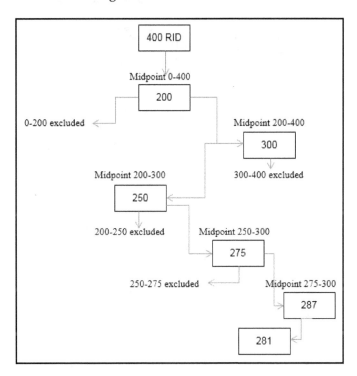

We start off from 400 observations and keep on calculating the midpoints until we reach a subset of the population that is small enough to help us identify our 281st RID.

In terms of the storage of the index, we know that the SAS data is stored in pages, while indexes are stored in ascending order. Let's look at how the index is stored for our AC dataset:

```
Proc Contents Data=Work.AC Centiles;
Run;
```

This is the resultant output showing the centiles for index:

	Alphabetic List of Indexes and Attributes				
#	Index	Update Centiles	Current Update Percent	# of Unique Values	Variables
1	City	5	0	4	
					Adelaide
					Adelaide
					Adelaide
					Copenhagen
					Copenhagen
					Hong Kong
					Hong Kong
					Hong Kong
					Hong Kong
					Hong Kong
					Hong Kong
					Hong Kong
					Hong Kong
					Hong Kong
					hong Kong
					hong Kong
					hong Kong
					hong Kong
					hong Kong
					hong Kong
					hong Kong

In SAS, centiles are like cumulative percentiles. 21 values are stored in the index descriptor. They represent the 0-100th percentile at an interval of 5. In our example, we can see that Hong Kong has 9 percentiles. This is a high proportion of the data that is being occupied by one value of the indexed variable. Based on this, we can say that the index is not fairly distributed. Even if this dataset had millions of observations, given this distribution, it wouldn't be efficient to create an index on this variable.

This is because SAS will use the centile information to determine whether it is efficient to use the index to read the dataset or whether it directly read the dataset. In our case, there is less discrimination in the variable values and SAS may directly read the dataset. If the dataset changes by more than 5% of the index variable, then the centiles will be updated. This default of 5% can be altered to a user-specified percentage.

Unique values

We had four unique values in nine observations in the AC dataset. What if all the observations had unique values? Think about a customer ID level dataset containing the monthly summary of all credit and debit transactions of a retail bank. The aim of such a dataset is to produce a single record for each customer every month. But what if there were processing errors in the past and the customer ended up having multiple records in the summary file? Well, an index can come in handy – not only to retrieve records but also to ensure such a mistake is spotted.

Let's try to delete the index on the AC file and generate a new Unique index:

```
Proc Datasets Library = Summary;
Modify Summary;
   Index Delete City;
Run;
```

We will now try to create a Unique index:

```
PROC DATASETS LIBRARY=WORK;
MODIFY AC;
        Index Create City / Unique;
RUN;
```

In this instance, the Unique option would ensure that if there are duplicate records of the ID in the file, an error will be generated.

We get the following LOG message when we run the program:

```
ERROR: Duplicate values not allowed on index City for file AC.
 81 RUN;

 NOTE: Statements not processed because of errors noted above.
```

It's not possible to define a Unique index for the City variable as we have duplicate values. This demonstrates that you can use the Unique option to ensure that you don't have duplicate values where they are least expected.

Missing values

If there are a lot of missing values in the data, but it is still beneficial to create an index, use the `Nomiss` option. We will use the same dataset as used earlier in this chapter to create an index for `Housing`:

```
PROC DATASETS LIBRARY=WORK;
MODIFY AC;
        Index Create Housing / Nomiss;
RUN;
```

This will ensure that the index is created successfully but that the missing values are not added to it. This will ensure that a lot of resources and space are saved in creating the index.

Be careful that BY processing or the WHERE clause will not use the index if the missing values are qualified in these statements. The program specified will still execute, but the index file will not be leveraged:

```
Data Subset;
Set AC;
Where Housing LT 88;
Run;
```

However, in the following statement, the index would be leveraged:

```
Where Housing LT 88 or Housing NE .;
```

Now that you have learned about the various methods of combining, let's look into how to encrypt the data in a dataset.

Encryption

In Chapter 1, *Introduction to SAS Programming*, we encrypted a dataset using a password. Encryption can be done both at the Server and the SAS programming level. Most users will never get permission to encrypt at the Server level. However, SAS does offer two methods by which we can encrypt via programming. Earlier, we witnessed encryption via the built-in mechanism in SAS. The other method, known as **advanced encryption standard (AES)** - 256, is showcased in the following code block. 256 refers to the bit key for encryption.

The following is the syntax for SAS proprietary encryption:

```
Data Library.File (encrypt = yes pw = password);
Data Library.File (encrypt = yes write = password);
Data Library.File (encrypt = yes alter = password);
Data Library.File (encrypt = yes read = password);
```

The following is the syntax for AES encryption:

```
Data Library.File (encrypt = aes pw = key);
```

`key` can be without a quotation, or have a single or double quotation mark. The naming conventions for quotation marks are as follows.

No quotation marks:

- Use alphanumeric characters and underscores only
- Can be up to 64 bytes long
- Use uppercase and lowercase letters
- Must start with a letter
- Cannot include blank spaces
- Is not case-sensitive

Single quotation marks:

- Use alphanumeric, special, and DBCS characters
- Can be up to 64 bytes long
- Use uppercase and lowercase letters
- Can include blank spaces, but cannot contain all blanks
- Is case-sensitive

Double quotation marks:

- Use alphanumeric, special, and DBCS characters
- Can be up to 64 bytes long
- Use uppercase and lowercase letters
- Can include blank spaces, but cannot contain all blanks
- Is case-sensitive

Let's try both SAS proprietary and AES encryption:

```
Data Locked (Encrypt=Yes PW=TestKey);
Set AC;
Run;
```

The proprietary method is successful. After program execution, we get a dialog box where we need to enter the password specified to be able to see the output dataset:

In the following code block, we get an error in the SAS University Edition when trying to secure a file using AES encryption:

```
Data LockedAES (Encryptkey=AES PW=TestKey);
Set AC;
Run;
```

The error in the LOG states the following:

```
73 Data LockedAES (Encrypt=AES PW=XXXXXXX);
                         ___
                         301
   ERROR 301-63: SAS/SECURE is not licensed, but is required for strong
   encryption.

74 Set AC;
75 Run;
```

The SAS/Secure module is not available as part of SAS University Edition. However, users can still continue to use the proprietary encryption method in the University Edition.

Summary

In this chapter, we looked at various options for combining data. We explored concatenating datasets, interleaving them, and merging them. We also looked at altering and updating datasets. All the methodologies were compared in terms of their pros and cons.

We also looked at mastering combining datasets. When our database numbers increase, combining datasets is one thing we must do to keep everything neat and tidy. With combining datasets comes the problem of faster and more efficient ways to access datasets and protecting the data. Hence, we also looked at indexing and encrypting data. You will end up working with multiple datasets in real-world scenarios, as seldom is data held in one table. Indexing may be required to boost performance, whereas encryption is becoming the norm due to concerns about data security.

In the next chapter, we will learn about transforming procedures and functions, producing various statistics, and generating reports.

Power of Statistics, Reporting, Transforming Procedures, and Functions

4

When we introduced SAS, we primarily reviewed functions and looked at data merging. There are a host of built-in procedures that can help reduce coding efforts and provide flexibility so that we can transform data, produce statistics, run statistical tests (for a detailed explanation of some of these tests, please refer to *SAS for Finance* by Harish Gulati, available from Packt Publishing), and produce reports. In this chapter, the procedures we will explore are useful for reporting statistics and testing various hypotheses via statistical tests. The transpose function we will learn about in this chapter will help us to prepare data for various purposes, such as reporting and statistical tests.

In this chapter, we will cover the following procedures and functions:

- The following procedures will be covered:
    - Freq
    - Univariate
    - Means and summary
    - Corr
    - REG
- The following functions will be covered:
    - Transpose

Proc Freq

This procedure is primarily used to produce frequency counts. However, what distinguishes this procedure between others is its ability to compute Chi-Square tests, measures of association, and agreement for contingency tables. The syntax of the procedure is as follows:

```
Proc Freq <options>;
   Tables requests < options >;
   By variables;
   Exact statistic-options </ computation-options>;
   Output <OUT = SAS-dataset> options;
   Tables requests </options>;
Test options;
   Weight variable < / option >;
```

We will use the simplest form of version of `Proc Freq` and explore the results in a dataset containing information on the `Class`, `Height`, and `Weight` of children in a school:

```
Data Class;
Input ID Class $ Height $ Weight $;
Datalines;
1 A Over5.7 Above50 1 0 1
2 A Over5.7 Above50 1 1 0
3 B Over5.7 Below50 1 1 .
4 B Under5.7 Below50 1 1 1
5 A Over5.7 Below50 1 1 1
6 A Over5.7 Above50 1 . 1
;

Proc Freq Data = Customer_X;
Run;
```

A run of the basic `Proc Freq` helps us explore the dataset using some basic statistics:

		The FREQ Procedure		
ID	Frequency	Percent	Cumulative Frequency	Cumulative Percent
1	1	16.67	1	16.67
2	1	16.67	2	33.33
3	1	16.67	3	50.00
4	1	16.67	4	66.67
5	1	16.67	5	83.33
6	1	16.67	6	100.00

The output also contains the **Frequency**, **Cumulative Frequency**, **Percent**, and **Cumulative Percent** figures for all of the variables in our dataset:

Class	Frequency	Percent	Cumulative Frequency	Cumulative Percent
A	4	66.67	4	66.67
B	2	33.33	6	100.00

Height	Frequency	Percent	Cumulative Frequency	Cumulative Percent
Over5.7	5	83.33	5	83.33
Under5.7	1	16.67	6	100.00

Weight	Frequency	Percent	Cumulative Frequency	Cumulative Percent
Above50	3	50.00	3	50.00
Below50	3	50.00	6	100.00

We only have three variables in the current dataset. We might end up with many more variables in our dataset and the output can easily get out of control. Also, we might want to suppress some of the default output and add various other options. We will explore these options by specifying a separate statement:

```
Proc Freq Data = Class;
Table Class;
Run;
```

The addition of the `Table` statement will ensure that the output will only contain the frequency count, cumulative frequency, percentage, and cumulative percentage figures for the variables specified in the statement.

Cross tabulation

We have produced the frequency of each variable separately. `Proc Freq` can also produce cross tables or n-way tables, as follows:

```
Proc Freq Data = Class;
Table Class*Height;
Run;
```

This will result in the following table:

The FREQ Procedure			
Frequency Percent Row Pct Col Pct	Table of Class by Height		
		Height	
Class	Over5.7	Under5.7	Total
A	4 66.67 100.00 80.00	0 0.00 0.00 0.00	4 66.67
B	1 16.67 50.00 20.00	1 16.67 50.00 100.00	2 33.33
Total	5 83.33	1 16.67	6 100.00

We now have the frequency, percent, row, and column percentage statistics for the **Height** for each class. Let's describe each of these statistics:

- **Frequency**: This produces the frequencies for each column of the output. For **Over5.7**, there are **4** records corresponding to Class **A** and **1** record for Class **B**. In total, there are **5** records for **Over5.7**.
- **Percent**: This is the percentage of frequency in each column for the **Class**, divided by the total, that is, 4/5 for Class **A** and 1/5 for Class **B**, expressed as a percentage.
- **Row Percentage**: This is different from the previous statistics as it focuses on the row instead of the column. For Class **A**, we have **4** records for **Over5.7** and **0** records for **Under5.7**. Hence, the row percentage for cell (1,1) is **100.00**.
- **Col Percentage**: This is the opposite of row percentage. The metric for cell (1,1) is calculated by dividing the number of observations for **Over5.7** for Class **A** by the total number of observations for **Over5.7**.

Restricting Proc Freq output

On certain occasions, you may want to keep the output simpler. The following options will become even more useful when you have three-way cross tables, which we will explore shortly:

```
Proc Freq Data = Class;
Table Class*Height / nocol norow nocum nofreq;
Run;
```

	The FREQ Procedure			
Percent		Table of Class by Height		
		Height		
Class	Over5.7	Under5.7	Total
A	66.67	0.00	66.67
B	16.67	16.67	33.33
Total	5. 83.33	1 16.67	6 100.00

Cross tabulation with a controlling variable

The real use of Proc Freq comes when we produce n-way or cross-tables with more than two variables. In the following example, we introduce a three-way cross-tabulation by including the Weight variable. Notice that the output is segregated into two tables with the first specified variable, Class, being the common factor:

```
Proc Freq Data = Class;
Table Class*Weight*Height / nocol norow nocum nofreq;
Run;
```

This results in the following output:

The FREQ Procedure				

Percent	Table 1 of Weight by Height			
	Controlling for Class=A			

		Height		
Weight	Over5.7	Under5.7	Total	
Above50	75.00	0.00	75.00	
Below50	25.00	0.00	25.00	
Total	4 100.00	0 0.00	4 100.00	

Percent	Table 2 of Weight by Height			
	Controlling for Class=B			

		Height		
Weight	Over5.7	Under5.7	Total	
Above50	0.00	0.00	0.00	
Below50	50.00	50.00	100.00	
Total	1 50.00	1 50.00	2 100.00	

The order of the variables in the `Proc Freq` statement is important in determining the controlling variable for a cross-tabulation output. Look at how not putting an asterisk between the variables will change the output:

```
Proc Freq Data = Class;
Table Class Weight*Height / nocol norow nocum nofreq;
Run;
```

This will result in the following output:

The FREQ Procedure		

Class	Frequency	Percent
A	4	66.67
B	2	33.33

Percent	Table of Weight by Height			

		Height		
Weight	Over5.7	Under5.7	Total	
Above50	50.00	0.00	50.00	
Below50	33.33	16.67	50.00	
Total	5 83.33	1 16.67	6 100.00	

The **Class** variable has no interaction with the other variables. The **FREQ procedure** outputs the frequency of the variable but ignores it in the cross-tabulation output. Apart from the sequence of the variables, the asterisk is also important to the `Proc Freq` procedure.

Use the following program if you want to produce a frequency chart along with the tables. Remember to ensure that the ODS Graphics option is turned on:

```
Ods graphics on;
Proc Freq Data = Class;
Table Class Weight*Height / plots=FreqPlot;
Run;
```

The power of `Proc Freq` isn't restricted to producing frequency tables and charts. While this isn't a book about statistical models, let's see how the procedure can lend us help to analyze a scenario.

Proc Freq and statistical tests

We have data on injuries that have been sustained by people while playing various sports. For all of the activities, we know whether the individuals performed any warm-up exercises prior to playing. There is a suspicion that the lack of warm-ups led to higher injuries:

```
Data SportsInjury;
Input Activity $ Warmup Injury Count;
Datalines;
Running 1 0 5
Running 0 1 15
Running 1 1 3
Football 1 0 16
Football 1 1 10
Football 1 1 4
Squash 1 0 2
Squash 0 1 10
Squash 1 1 0
Weights 1 0 12
Weights 0 1 6
Weights 1 1 3
Others 0 0 10
;
```

We have aggregated the data at the level of the `Warmup` and `Injury` variables:

```
Proc Sql;
  Create table Analyse as
  Select Warmup, Injury, Sum(Count) as Cases
  From SportsInjury
  Group by 1,2;
Quit;

Proc Print Noobs;
Run;
```

This will give the following spot injury data output:

Warmup	Injury	Cases
0	0	10
0	1	31
1	0	35
1	1	20

We have four possible scenarios, as follows:

- No warm-up is done, leading to no injury (200 cases)
- No warm-up is done, leading to injury (31 cases)
- A warm-up is done, leading to no injury (35 cases)
- A warm-up is done, leading to injury (20 cases)

We will now try and see whether there is any statistical significance between not doing warm-ups and having a sports injury. We will run a Chi-Square test and extract measures of a Relative Risk test, as follows:

```
Proc Freq Data=Analyse;
Tables Warmup*Injury / chisq relrisk;
Weight Cases;
Run;
```

The first table in the output contains a cross-tabulation of the `Warmup` and `Injury` variables:

Frequency Percent Row Pct Col Pct	Table of Warmup by Injury		
		Injury	
Warmup	0	1	Total
0	10 10.42 24.39 22.22	31 32.29 75.61 60.78	41 42.71
1	35 36.46 63.64 77.78	20 20.83 36.36 39.22	55 57.29
Total	45 46.88	51 53.13	96 100.00

This table is similar to the output we saw in our previous examples.

The Chi-Square statistic is significant, as shown here:

Statistics for Table of Warmup by Injury			
Statistic	DF	Value	Prob
Chi-Square	1	14.5288	0.0001
Likelihood Ratio Chi-Square	1	15.0520	0.0001
Continuity Adj. Chi-Square	1	12.9955	0.0003
Mantel-Haenszel Chi-Square	1	14.3774	0.0001
Phi Coefficient		-0.3890	
Contingency Coefficient		0.3626	
Cramer's V		-0.3890	

We need to test the relation of warm-ups and injury by using Fisher's Exact Test. The output in the cell (1,1) in the **Table of Warmup by Injury** corresponds to the order of the scenarios we have in the previous section. The scenario implies that, when there was no warm-up, there was no injury in 10 of the cases. However, this isn't a scenario that we want to test. A scenario that leads to injury with or without the warm-up being undertaken would help us test our assumption of the role of warm-ups more efficiently. Because Fisher's Exact Test only tests for cell (1,1), we need to ensure our data is properly sorted.

We need to alter the scenario that's tested in the following data:

Fisher's Exact Test	
Cell (1,1) Frequency (F)	10
Left-sided Pr <= F	0.0001
Right-sided Pr >= F	1.0000
Table Probability (P)	0.0001
Two-sided Pr <= P	0.0002

We can achieve this by sorting the data in the desired order and then ensuring that `Proc Freq` retains the sorted order. We retain the desired order by adding the `Order=Data` option to the `Proc Freq` statement:

```
Proc Sort Data = Analyse;
By Warmup descending Injury;
Run;

Proc Freq Data=Analyse Order=Data;
Tables Warmup*Injury / chisq relrisk;
Weight Cases;
Run;
```

The Chi-Square test statistic isn't affected by the change in order. However, the output from Fisher's Exact Test and the Odds Ratio and Relative Risk is altered, owing to the sorting of the data. These two aspects are changed because the `Proc Freq` cross-tabulation is altered. The `Proc Freq` output is mentioned so as to help you understand the hypothesis we will be testing in the Fisher's Exact Test:

Frequency Percent Row Pct Col Pct	Table of Warmup by Injury		
		Injury	
Warmup	1	0	Total
0	31 32.29 75.61 60.78	10 10.42 24.39 22.22	41 42.71
1	20 20.83 36.36 39.22	35 36.46 63.64 77.78	55 57.29
Total	51 53.13	45 46.88	96 100.00

We will be testing for instances of injury when no warm-up has been undertaken. This is represented by the left-sided test in the following screenshot. Our tests are as follows:

- **Two-tailed**: The odds of injury differ between warming up or not warming up.
- **Left tailed**: The odds of injury are less in instances of no warm-up.
- **Right tailed**: The odds of injury are more in instances of no warm-up.

The right-tailed test seems best suited for our assumption. Since the p-value is significant, we accept the hypothesis that there is a greater risk of injury in instances of no warm-up:

Fisher's Exact Test	
Cell (1,1) Frequency (F)	31
Left-sided Pr <= F	1.0000
Right-sided Pr >= F	0.0001
Table Probability (P)	0.0001
Two-sided Pr <= P	0.0002

The odds ratio provides an estimate of the Relative Risk when an event is rare. The estimate indicates that the odds of injury are 5.43 times higher in the no warm-up instance:

Odds Ratio and Relative Risks			
Statistic	Value	95% Confidence Limits	
Odds Ratio	5.4250	2.2058	13.3426
Relative Risk (Column 1)	2.0793	1.4071	3.0724
Relative Risk (Column 2)	0.3833	0.2157	0.6810
Sample Size = 96			

However, the wide confidence limit indicates that the estimate has low precision.

Proc Univariate

This is a procedure that can accomplish many tasks. However, it is frequently used for producing descriptive statistics based on moments (including skewness and kurtosis), quantiles or percentiles, frequency tables, and extreme values. It can produce a host of charts and goodness of fit tests for a lot of distribution types.

Since the previous function dealt with frequency tables, let's kick-start learning about this procedure by producing something similar. Use the following code to produce the frequency table:

```
ODS Select Frequencies;
Proc Univariate Data = Analyse Freq;
Var _All_;
Run;
```

This will result in the following output:

The UNIVARIATE Procedure
Variable: Warmup

Frequency Counts

		Percents	
Value	Count	Cell	Cum
0	2	50.0	50.0
1	2	50.0	100.0

The UNIVARIATE Procedure
Variable: Injury

Frequency Counts

		Percents	
Value	Count	Cell	Cum
0	2	50.0	50.0
1	2	50.0	100.0

The UNIVARIATE Procedure
Variable: Cases

Frequency Counts

		Percents	
Value	Count	Cell	Cum
10	1	25.0	25.0
20	1	25.0	50.0
31	1	25.0	75.0
35	1	25.0	100.0

As you can see, we can produce frequencies using `Proc Univariate`. However, if frequencies are the main analysis goal, try and use `Proc Freq`. If you omit the ODS Select option at the top of the query, you will get a detailed output. The **Output Delivery System (ODS)** stores various outputs. By using frequencies, we have requested the ODS to only output the table that's saved under this particular name.

Basic statistics and extreme observations

The following table shows the credit transactions of a few customers. We will produce an output with basic measures and identify extreme observations:

```
Data Transactions;
   Input CustId $ Credit;
Datalines;
A2112 234
A2342 532
A2345 345
A6345 234
B3234 234
B6345 456
C465A 675
D4436 790
E4603 645
F0945 709
F435F 999
H0032 009
;
```

Once again, we have restricted the output using the ODS facility to ensure only part of the output is written out:

```
ODS Select BasicMeasures Extremeobs;
Proc Univariate Data = Transactions;
Var Credit;
Run;
```

If we have multiple variables, we can request basic statistics by adding them in the `Var` statement:

The UNIVARIATE Procedure Variable: Credit			
Basic Statistical Measures			
Location		Variability	
Mean	488.5000	Std Deviation	287.17733
Median	494.0000	Variance	82471
Mode	234.0000	Range	990.00000
		Interquartile Range	458.00000

The extreme observations are segregated into the lowest and highest in terms of value and observations:

Extreme Observations			
Lowest		Highest	
Value	Obs	Value	Obs
9	12	645	9
234	5	675	7
234	4	709	10
234	1	790	8
345	3	999	11

Here, observations **11** and **12** represent outliers.

Tests of normality

A normal distribution is a probability function that describes how the values are distributed. Most of the observations are centered and the probability for values that lie at the extreme is equally spread on both sides. A normal distribution is also described as a bell curve:

```
Ods Graphics Off;
Proc Univariate Data = Transactions;
Histogram Credit / normal name='MyPlot';
Inset n normal(ksdpval) / pos = ne Format = 5.3;
Run;
```

The following is a partial output of the Univariate Procedure:

The UNIVARIATE Procedure Variable: Credit			
Moments			
N	12	Sum Weights	12
Mean	488.5	Sum Observations	5862
Std Deviation	287.177329	Variance	82470.8182
Skewness	0.0869798	Kurtosis	-0.6454251
Uncorrected SS	3770766	Corrected SS	907179
Coeff Variation	58.7875801	Std Error Mean	82.900954

We have created a histogram and drawn a normal distribution plot:

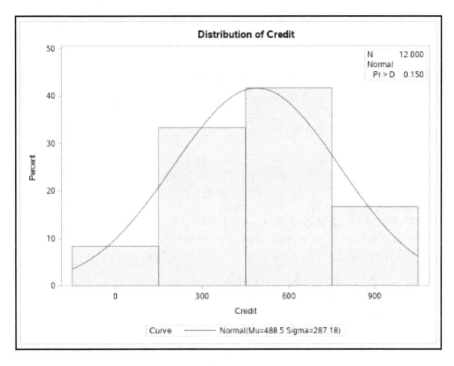

The tests for location have a small p-value:

Tests for Location: Mu0=0				
Test		Statistic	p Value	
Student's t	t	5.892574	Pr > |t|	0.0001
Sign	M	6	Pr >= |M|	0.0005
Signed Rank	S	39	Pr >= |S|	0.0005

The null hypothesis that credit is not normally distributed is rejected due to a p-value greater than 0.10 for the **Kolmogorov-Smirnov (KS)** test:

Goodness-of-Fit Tests for Normal Distribution				
Test		Statistic	p Value	
Kolmogorov-Smirnov	D	0.14558169	Pr > D	>0.150
Cramer-von Mises	W-Sq	0.03543167	Pr > W-Sq	>0.250
Anderson-Darling	A-Sq	0.22494717	Pr > A-Sq	>0.250

Tests for location

You may have a business problem where certain stakeholders might assume a customer lifetime value to be equal to a certain number. You can use `Proc Univariate` to test the is hypothesis. In our example, let's assume that the mean of credit transactions is 200. This is our hypothesis. We will run three tests of location to accept or reject the hypothesis:

```
ODS Select TestsforLocation LocationCounts;
Proc Univariate Data = Transactions MU0=200 Loccount;
Var Credit;
Run;
```

All three tests have a small p-value of less than 0.05, and hence we reject the null hypothesis that the mean of the distribution is 200. Users of such tests need to be aware that at a hypothesized mean that's equal to 300, you will fail to reject the null hypothesis. However, as seen in the **UNIVARIATE Procedure** table in the previous section, the mean of the observations is 488.5. Hence, the inability to reject the null hypothesis might make you believe that 300 is representative of the real mean of the observations:

The UNIVARIATE Procedure
Variable: Credit

Tests for Location: Mu0=200

Test		Statistic		p Value	
Student's t	t	3.480056	Pr > |t|	0.0051
Sign	M	5	Pr >= |M|	0.0063
Signed Rank	S	34	Pr >= |S|	0.0044

We have requested for a count of the observations that are greater than, not equal to, and less than 200:

Location Counts: Mu0=200.00

Count	Value
Num Obs > Mu0	11
Num Obs ^= Mu0	12
Num Obs < Mu0	1

Proc Means and Summary

In the earlier versions of SAS, Proc Means and Summary had distinct features. Over the last few versions, the only difference between the two procedures are as follows:

- Proc Means outputs the results in the listing window or in any other open output destination, whereas Proc Summary creates a dataset by default. You cannot execute a Proc Summary without an output statement.
- In the absence of a `Var` statement, Proc Means analyzes all numeric variables. Proc Summary, in the same situation, only produces a count of the observations.

Proc Means

The simplest form of `Proc Means` is as follows:

```
Proc Means Data = Transactions;
Run;
```

This will give us the following output:

The MEANS Procedure				
Analysis Variable : Credit				
N	Mean	Std Dev	Minimum	Maximum
12	488.5000000	287.1773288	9.0000000	999.0000000

We can also add a `Class` variable to the analysis:

```
Proc Means Data = SportsInjury;
Class Warmup Injury;
Var Count;
Run;
```

This will result in the following output:

				The MEANS Procedure				
				Analysis Variable : Count				
Warmup	Injury	N Obs	N	Mean	Std Dev	Minimum	Maximum
0	0	1	1	10.0000000	.	10.0000000	10.0000000
	1	3	3	10.3333333	4.5092498	6.0000000	15.0000000
1	0	4	4	8.7500000	6.3966137	2.0000000	16.0000000
	1	5	5	4.0000000	3.6742346	0	10.0000000

The addition of a BY statement is possible. However, the data needs to be sorted according to the BY variables:

```
Proc Sort Data = Customer_X;
By Class;
Run;

Proc Means Data = Customer_X;
By Class;
Class Height;
Var Basketball;
Run;
```

You may notice similarities between the outputs of the `Freq`, `Univariate`, and `Means` procedures when basic statistics are being reported. The following is the resultant output of `Proc Means` By:

			The MEANS Procedure				
			Class=A				
			Analysis Variable : Basketball				
Height	N Obs	N	Mean	Std Dev	Minimum	Maximum
Over5.7	4	3	0.6666667	0.5773503	0	1.0000000

			Class=B				
			Analysis Variable : Basketball				
Height	N Obs	N	Mean	Std Dev	Minimum	Maximum
Over5.7	1	1	1.0000000	.	1.0000000	1.0000000
Under5.7	1	1	1.0000000	.	1.0000000	1.0000000

Proc Summary

Let's produce a similar output to Proc Means using the same `By` and `Class` groups:

```
Proc Summary Data = Customer_X;
By Class;
Class Height;
Var Basketball;
Output Out=Test N=n Mean=mean STD=stdev Min=min Max=max;
Run;

Proc Print Data=_LAST_ (Drop = _TYPE_ Rename=(_FREQ_=Nobs))
Noobs;
Where Height ne "";
Run;
```

While printing this out, we have tweaked the output to align it with the output of Proc Means By:

Class	Height	Nobs	n	mean	stdev	min	max
A	Over5.7	4	3	0.66667	0.57735	0	1
B	Over5.7	1	1	1.00000	.	1	1
B	Under5.7	1	1	1.00000	.	1	1

Proc Corr

Correlation can be defined as the linear relationship between multiple variables. Positive, negative, and no correlation are the three scenarios that can exist. The values range from -1 to 1. A positive correlation is observed when the value of one variable increases with the increase of another variable, whereas in negative correlation, with an increase in the value of one variable, the value of another variable decreases.

Correlation is often run as a precursor to evaluating the role of variables in models such as regression. In the following instance, we have multiple variables in the `Var` statement that are being assessed as predictors for the `Stock` variable. The `Date` variable is an ID variable and is not being assessed:

```
Proc Corr Data = Model;
ID Date;
With Stock;
Var Basket_Index -- M1_Money_Supply_Index;
Run;
```

The Corr procedure produces basic statistical measures similar to the procedure outputs we saw earlier in this chapter:

Simple Statistics							
Variable	N	Mean	Std Dev	Sum	Minimum	Maximum	Label
Stock	594	4.89662	0.64370	2909	3.43000	5.96000	Stock
Basket_index	594	152.26094	7.26174	90443	139.00000	169.00000	Basket_index
EPS	594	3.75315	0.17546	2229	3.45000	3.97000	EPS
Top_10_GDP	594	2.82684	0.36737	1679	2.10000	3.23000	Top_10_GDP
Global_mkt_share	594	0.19217	0.00136	114.14990	0.18850	0.19330	Global_mkt_share
P_E_ratio	594	18.09549	1.45594	10749	15.23000	20.24000	P_E_ratio
Media_analytics_index	594	196.52357	5.82027	116735	185.00000	209.00000	Media_analytics_index
Top_10_Economy_inflation	594	2.54293	0.21919	1511	2.25000	2.86000	Top_10_Economy_inflation
M1_money_supply_index	594	119.51347	4.47221	70991	112.00000	125.00000	M1_money_supply_index

As you can see, some of the variables have a positive high correlation (tending to 1) and that one variable has a high negative correlation (tending to -1):

	Basket_index	EPS	Top_10_GDP	Global_mkt_share	P_E_ratio	Media_analytics_index	Top_10_Economy_inflation	M1_money_supply_index
	Pearson Correlation Coefficients, N = 594 Prob > |r| under H0: Rho=0							
Stock	0.73477	0.84048	0.74627	0.66585	0.58033	0.48898	0.87827	-0.84928
Stock	<.0001	<.0001	<.0001	<.0001	<.0001	<.0001	<.0001	<.0001

As part of the current code, we have only tested for correlation and not significance.

Proc REG

We will take this example forward and test the significance of the variables in a regression model:

```
Data Build Validation;
Set Model;
If Date lt '01Dec2017'd then output Build;
Else output Validation;
Run;

PROC REG DATA=build plots=diagnostics(unpack);
ID date;
MODEL stock = basket_index -- m1_money_supply_index;
RUN;
```

The observations that have been used have decreased in the regression model. Previously, in the correlation procedure, we had **594** rows of data, which has now decreased to **564**. We have called the new data the build data. The last **30** observations in the data have been left out of the model building process and have been put in a dataset called **validation**:

The REG Procedure
Model: MODEL1
Dependent Variable: Stock Stock

Number of Observations Read	564
Number of Observations Used	564

Analysis of Variance					
Source	DF	Sum of Squares	Mean Square	F Value	Pr > F
Model	8	188.76933	23.59617	540.40	<.0001
Error	555	24.23354	0.04366		
Corrected Total	563	213.00287			

In the **Analysis of Variance** (ANOVA) table, the eight degrees of freedom refers to the eight independent variables that are available to estimate the parameters of predicting the dependent variable. Total degrees of freedom represent the sources of variance, which is usually denoted by *N-1*. Since we did not exclude the intercept from the modeling statement, we have *N-1 (8 independent variables - 1) + intercept*, which gives us 8 degrees of freedom.

The error that's mentioned in the ANOVA table is something that most modelers actively don't consider while assessing the results of the regression. However, if the error is equal to 0, then no F or P statistic will be produced in ANOVA and the modeler will probably have to collect more data before the regression model can be run. The error term refers to the residual degrees of freedom which, in our case, will be *563-8 = 555*.

The sum of squares in ANOVA is associated with the variance for the model, error, and the corrected total. The mean square refers to the sum of squares divided by the degrees of freedom. The F value is compiled by dividing the Mean Square Model by the Mean Square Error. The F value, when used in conjunction with the P-value, informs us whether the independent variables are reliable variables for the prediction of the dependent variable stock.

The next important bit of statistics that the modeler will refer to is the values of R-Square and Adjusted R-Square, as shown in the following screenshot. The R-Square is the value that depicts the percentage of the variance in the variable stock that can be explained by the independent variable. In this case, 88.59% percent of the variance can be explained by the use of the independent variables that have been selected for modeling. As the independent variables keep getting added on to the model, the prediction power of the model increases in general. Some of this increase happens by chance. To negate this chance factor, we observe the Adjusted R-Square. A close R-Square and Adjusted R-Square is always a good sign in the model. One of the reasons that the Adjusted R-Square could vary significantly from the R-Square is the low number of observations and a high number of predictor variables. Apart from this, if the two values differ significantly, then the modeler should ensure that the model is thoroughly scrutinized. A higher Adjusted R-Square value is a positive sign in interpreting the model, but it isn't the only measure that the modeler should rely on.

The parameter estimate table is critical for the model-building exercise as this is the section that helps identify the significant variables for prediction and informs us of the value of the parameter estimate that is to be used in the regression equation. However, some modelers tend to overlook the observations that can be made from looking at the diagnostic plots. Before accepting the results from the model, it is necessary to review the diagnostic plots. But for now, let's focus on the value of the information in the parameter estimates table:

Root MSE	0.20896	R-Square	0.8862
Dependent Mean	4.84254	Adj R-Sq	0.8846
Coeff Var	4.31508		

Parameter Estimates							
Variable	Label	DF	Parameter Estimate	Standard Error	t Value	Pr > |t|
Intercept	Intercept	1	-3.02887	4.47183	-0.68	0.4985
Basket_index	Basket_index	1	-0.00550	0.00237	-2.32	0.0206
EPS	EPS	1	1.62607	0.16143	10.07	<.0001
Top_10_GDP	Top_10_GDP	1	0.30274	0.11734	2.58	0.0101
Global_mkt_share	Global_mkt_share	1	102.74845	23.89044	4.30	<.0001
P_E_ratio	P_E_ratio	1	-0.12198	0.01065	-11.46	<.0001
Media_analytics_index	Media_analytics_index	1	0.01452	0.00223	6.51	<.0001
Top_10_Economy_inflation	Top_10_Economy_inflation	1	-1.26989	0.29079	-4.37	<.0001
M1_money_supply_index	M1_money_supply_index	1	-0.12856	0.00805	-15.96	<.0001

There are six variables according to the table, which are statistically significant variables in predicting the variable stock. All of these variables have *Pr> |t| <0.001*. We can set our significance level (alpha level) at 95% or 0.05 to accept or reject the null hypothesis. All six significant variables have a p-value of less than 0.05. The null hypothesis that's being tested here is that the explanatory variables don't have a significant explanatory power (as the regression coefficient is assumed to be zero) for the response variable of the stock price. In the case of these six variables, as the p-value is less than 0.05 (our significance level), we can reject the null hypothesis and conclude that these variables can be used for explaining the relationship with stock price movement.

For every change in one unit of the significant variable of EPS, there is a 1.62607 unit change in the value of the stock price. This level of change per unit is provided by the parameter estimate. The parameter estimate also provides the direction of the relationship. Three of the six significant variables are inversely related to the stock price. These are P/E ratio, the weighted inflation of the top 10 economies, and the weighted index of the M1 money supply for the top 10 economies. The modeler had expected that the P/E ratio could be inversely related to the stock price as a lower P/E ratio is what investors are typically looking for, with the expectation that investing in such a stock would lead to higher returns. Hence, the negative relationship between the P/E ratio and stock price makes sense.

If you can recall the correlation output, then you might have noticed that only the M1 money supply index variable had an inverse relationship when we measured the strength between each variable and stock price. However, in the regression model output, we have three significant variables with an inverse relationship with the stock price. Remember that we also said that correlation doesn't test the significance, and we do not assume any independent or dependent variable relationship in correlation. Furthermore, the correlation was just a measure between the two variables. It didn't take into account the effect of other variables in measuring the strength of the relationship among two variables, whereas, in the regression model, all independent variables are collectively trying to explain the variance that can be seen in the stock price movement. This interplay between various independent variables is different from the correlation phenomenon. Hence, the output of correlation and regression should be viewed in the correct perspective.

So far, we have analyzed the output of ANOVA and studied the relevance of the parameter estimates. We have also highlighted how no statistic should be interpreted in isolation and that diagnostic plots should also be evaluated. Let's move on to assessing the diagnostic plots.

The residual fit spread for variable stock shows two charts boxed together. These are the **Fit-Mean** and the **Residual**:

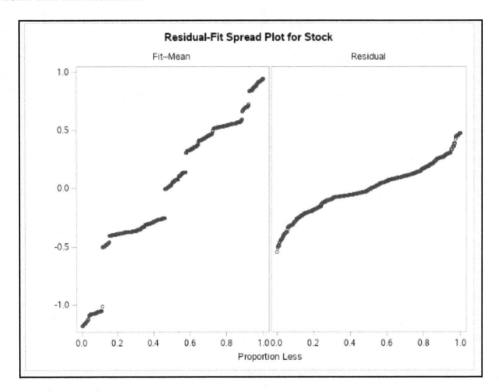

The **Fit-Mean** refers to the spread of the fitted values, while the Residual graph refers to the spread of the residuals. The residuals are calculated based on the difference between the observed (actual) and the fitted (predicted). In the case of our model, the spread of the fitted is more than the spread of the residual, which means that the spread of residual is less than the fitted and the model can be used. The flatter or more horizontal the shape of the residuals, the better the chances of the spread of fitted being more favorably distributed.

The Q-Q plot of residuals does point to some observations at the lower and top end of the quantile that have higher residual and haven't fit as well as the rest of the population:

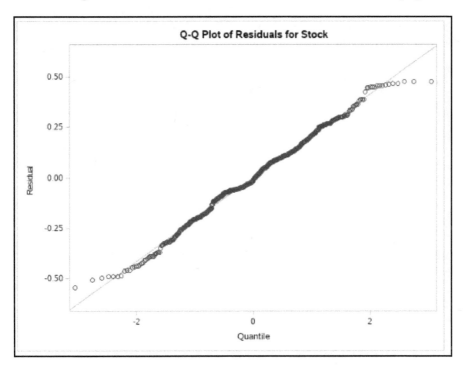

When the stock price is low, then the model is under-predicting the stock value in some cases and the stock price is at the higher end, and the model is over-predicting the value in some cases. This is an observation that the modeler noted down in case this bias can be overcome by building an alternative model.

The distribution of residuals is fairly normally distributed, with no particular skewness observed. The residuals by predicted chart show that residuals of predicted observations are higher when the stock price is around $4.80-4.90. This chart also reaffirms the observation we made earlier regarding that the model is over-predicting the stock price when dealing with higher observed stock prices. Most of the residuals beyond the $5.50 stock price are negative.

It is important for the modeler to decide whether the spread of the residuals is acceptable. This is a subjective call at times, and the level of acceptance around the spread differs from one business problem to another. One of the main assumptions of regression is that the residuals are normally distributed. By looking at the Q-Q plot, we can say that this assumption holds in this case:

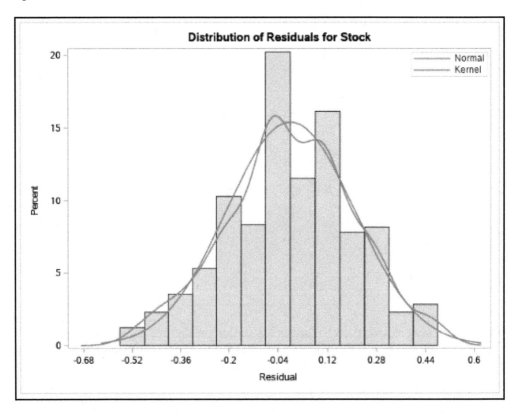

Another main assumption of regression is that the residuals shouldn't form any pattern. Looking at the residuals that have been predicted for the stock chart, it seems that the residuals do not form any particular patterns. Hence, this assumption about regression also holds true for the model:

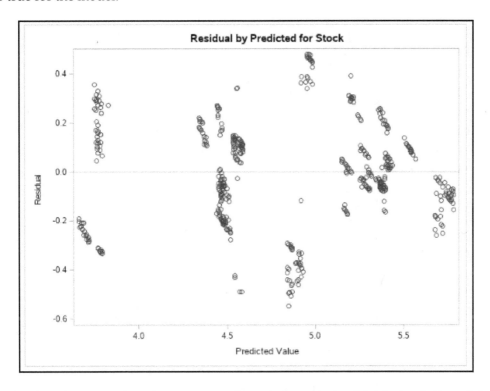

In this book, only some of the key aspects of Proc Reg are being highlighted. One of the aspects that's worth mentioning to users is the ability to generate scores for a holdout sample. The hold out sample contains one month of observed data. At times, while building the model, we overfit the data. It is always a good idea to have a holdout sample on which the model can be fitted. There isn't anything better than using the observed data to see how well the predictions were made. The holdout sample shouldn't be too short or too future-looking compared to what has been built in the model. We cannot expect the model that we have built for daily stock prices using almost three years of data to help us predict three years ahead. Also, using our model to just predict one day ahead and using that as a holdout sample would be too lenient on the model and would not check for its practical use with a decent data size.

We will write the model parameters to the REGOUT dataset. We have stored our hold out data in a table called validation. The regression equation will be used to generate forecasts on this dataset:

```
PROC SCORE Data=validation Score=REGOUT Out=RSCOREP
Type=PARMS;
Var basket_index eps p_e_ratio global_mkt_share media_analytics_index
m1_money_supply_index top_10_gdp;
RUN;

Proc Print Data = RSCOREP (Keep = Date Stock Model1
Rename = (Model1=Predicted_Stock_Value));
Run;
```

The partial print of the RSCOREP table showcases how we have managed to use the scores to produce a predicted value of the stock price for our holdout/validation sample:

Obs	Date	Stock	Predicted_Stock_Value
1	12/01/2017	5.85	5.82231
2	12/02/2017	5.85	5.78811
3	12/03/2017	5.85	5.78811
4	12/04/2017	5.86	5.80179
5	12/05/2017	5.87	5.83599
6	12/06/2017	5.87	5.80179
7	12/07/2017	5.87	5.80179
8	12/08/2017	5.87	5.77443
9	12/09/2017	5.87	5.81547
10	12/10/2017	5.88	5.79495

Proc Transpose

We have seen how to utilize some powerful procedures for statistical analysis. As a data user, the transformation of data from horizontal to vertical or regrouping between columns and rows is an important tactical step. This step could be necessary for forming the input to the modeling dataset or as an output to produce a report or showcase insights. You may want to transpose all the variables or just some of them. This is also an effective way to present variables in a grouped manner, without having to perform any mathematical aggregation.

We will use the variables from the following dataset to learn about transposing:

```
Data Base;
Input CustID Year Avg_Credit Avg_Debit Spend_Indicator $;
Datalines;
1010 16 235 245 R
1010 17 230 220 A
1010 18 235 200 G
1010 19 254 220 G
1011 16 653 650 A
1011 17 650 610 G
1011 18 640 620 G
1011 19 650 656 A
1012 16 569 569 R
1012 17 560 550 G
1012 18 550 550 R
1012 19 450 400 G
;
```

We will retain only two variables initially and run Proc Transpose without any arguments:

```
Data Base_Narrow (Keep = CustID Year);
Set Base;
Where CustID=1010;
Run;
```

This will result in the following output:

Obs	CustID	Year
1	1010	16
2	1010	17
3	1010	18
4	1010	19

Now, we will invoke Proc Transpose with on argument, where we will create a new output dataset. At this stage, we don't want to overwrite our input dataset:

```
Proc Transpose Data = Base_Narrow Out=Wide;
Run;
```

This results in the following output:

Obs	_NAME_	COL1	COL2	COL3	COL4
1	CustID	1010	1010	1010	1010
2	Year	16	17	18	19

This procedure has transformed the two variables in the dataset from columns into rows. Remember, Proc Transpose will do default transformation in the absence of any argument, but only for numeric variables.

Let's add another variable from the `Base` dataset and a few arguments to Proc Transpose:

```
Data Base_Narrow (Keep = CustID Year Avg_Credit);
Set Base;
Run;
```

This results in the following output:

CustID	Year	Avg_Credit
1010	16	235
1010	17	230
1010	18	235
1010	19	254
1011	16	653
1011	17	650
1011	18	640
1011	19	650
1012	16	569
1012	17	560
1012	18	550
1012	19	450

The challenge with multiple variables is not just about how to transform the variables. The bigger challenge is to visualize what the reshaped data should look like. The output should make intuitive sense and should be in a form that can be easily understood and analyzed for insights or used as input for further statistical analysis:

```
Proc Transpose Data = Base_Narrow Out=Wide;
By CustID;
ID Year;
Var Avg_Credit;
Run;
```

This results in the following output:

CustID	_NAME_	_16	_17	_18	_19
1010	Avg_Credit	235	230	235	254
1011	Avg_Credit	653	650	640	650
1012	Avg_Credit	569	560	550	450

The BY variable, as we have seen, acts as a grouping variable. Our data is already pre-sorted as per the BY variable. The ID variable used to specify a variable whose formatted values name the transposed variables. The ID variable cannot have duplicate values. If it does have duplicate values, use the LET argument in the DATA statement to force Proc Transpose to ignore the duplicates. This will prevent the procedure from showing a WARNING and stopping the execution. Instead, you will be presented with a WARNING but the values will be transposed. The VAR statement specifies the variable that is being transposed. The auto-generated _NAME_ variable specifies the name of the variable that has been transposed.

We have achieved the objective of Transpose. However, the variable naming procedure isn't intuitive for the newly formed columns. Also, the _NAME_ variable is repetitive. I can Alternatively, I can present the data as follows:

```
Proc Transpose Data = Base_Narrow Out=Wide (Drop =_NAME_)
     Prefix=Year;
By CustID;
ID Year;
Var Avg_Credit;
Run;

Title Height = 8pt "Average Credit of Customers Across Years";
Proc Print Noobs;
Run;
```

This will result in the following output:

Average Credit of Customers Across Years				
CustID	Year16	Year17	Year18	Year19
1010	235	230	235	254
1011	653	650	640	650
1012	569	560	550	450

This table is the Proc Transpose output with Prefix.

Summary

In this chapter, we primarily conducted a statistical analysis using some powerful built-in procedures. We also had a brief look at how data can be transposed as this is an important step prior to getting the data ready for statistical analysis or report presentation. We also explored how some of the outputs from these procedures can be used to achieve a common objective, such as producing frequencies. The mix of procedures we have used helped us to produce basic statistics. We also managed to build a model using the REG procedure. While learning to model isn't the goal of this book, the examples we've used can help us perform the necessary data transformations and code the model.

In the next chapter, we will learn about advanced programming techniques, which are also known as macros in SAS.

3

Section 3: Advanced Programming

This part assists in programming at an advanced level, where readers are introduced to a number of functions that facilitate the simplification of complex programming tasks. The concept of macros is introduced, enabling the reader to write their first macro.

This section comprises the following chapters:

- Chapter 5, *Advanced Programming Techniques: SAS Macros*
- Chapter 6, *Powerful Functions, Options, and Automatic Variables Simplified*

5
Advanced Programming Techniques - SAS Macros

In the earlier chapters, we learned about loops. As we know, they can help us do repetitive tasks in a faster and more efficient way. But macros go a step further. They help to do the following:

- Reuse code
- Specify a value in one instance but use the value in multiple instances
- Perform mathematical and logical operations and let the macro consume this information by itself

So should you always write macros? No, writing macros can get challenging. The more you want the macro to do, the more complex the coding process can get.

The following topics will be covered in this chapter:

- Introduction to macros and macro definitions
- Macro variable processing
- Macro resolution tracking
- Comparing positional and keyword parameters
- Data-driven programming
- Writing efficient macros

What are macros?

In common language, we call the macro facility *macros*. The macro facility is a means for text substitution. It has two main components:

- **Macro processor**: The portion of the facility that does the work
- **Macro language**: The syntax that communicates with the processor

Macros are generally used as macro variables or macro definitions. In the former, one or multiple variables can take the value from user inputs, or can be based on the mathematical or logical resolution of some statements. The macro definitions can consist of various tasks that can be called upon by the user.

Remember these *don'ts* even before you start writing macros, or else you might get caught up in programming rather than delivering your objectives efficiently:

- Macros can't fully substitute for usual programming. Some of the tasks you do in programming are non-repetitive and could be written in a simple manner without the use of macros. Avoid macros in such scenarios.
- Don't try to build a "solve everything" macro. This might not be a business requirement. At times, analysts want to build such a macro and keep it for future tasks or reference. Do weigh in the time required for building such a macro with task objectives. One major downside of building such a macro if you are employed is that you may not have the copyright for that macro and when you leave the job, you may not be entitled to carry the piece of code with you.
- Macros can be difficult to debug. At times, rather than reusing someone's macro, it might be easier to write your own.
- Break the macro into sub-macros. You don't have to put all the functionality in one macro. You can write multiple macros that do various tasks.
- The objective you want to achieve can be easily met by using one of the SAS functions. Essentially, SAS functions are prewritten macros that are packaged together for the end user. Always explore whether a function can deliver what the macro intends to do or at least aid in the macro operation.

Macro variable processing

Let's look at how macro processing happens. We did explore table creation in `Chapter 1`, *Introduction to SAS Programming*, where we discussed the compiler and execution phases. Macro processing is related to the compiler phase. The whole macro processing starts from the input buffer stage that we discussed earlier. The one aspect we didn't discuss in detail earlier was the word scanner. The word scanner is a component that reviews the characters from the input buffer and segregates them into tokens. Tokens are like atoms, the smallest pieces of information that can be held in the SAS processing engine. The process of breaking this information is called **tokenization**. The word scanner determines which part of SAS processing each token should be sent to.

The types of tokens are as follows:

- **Literals**: A string of characters enclosed in single or double quotation marks.
- **Names**: A string of characters beginning with an underscore or letter and continuing with letters. It can end with an underscore.
- **Numbers**: Digits or date/time values and hexadecimal numbers.
- **Special**: Characters other than letters, numbers, or underscores that have intuitive meaning to the SAS system.

We will examine the tokenization of the following program:

```
%Let File = Class;
Data New;
Set &File;
New_Weight=Weight*1.1;
Run;
```

In the following description of tokenization, you will come across the *symbol table*. This is a table that SAS creates at the start of each session to hold the automatic and global macro variables in its memory. The value of the macro variable we assigned earlier will be held in the symbol table for the duration of the SAS session.

The following diagram shows the SAS macro variable processing steps:

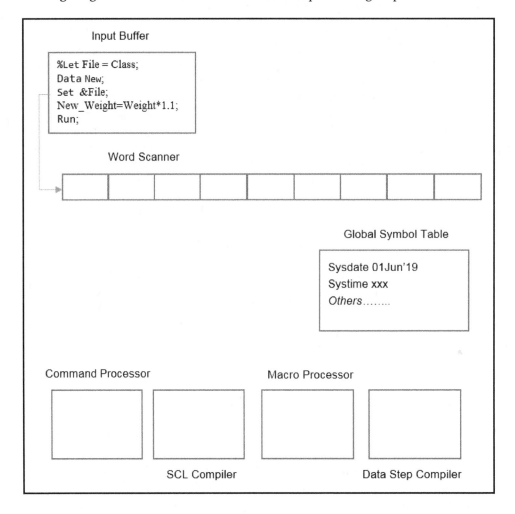

In the preceding diagram, we have not initiated the action of the **Word Scanner**. We will not be discussing the **Command Processor** and **SCL Compiler** as part of this chapter. The **Data Step Compiler** has already been discussed in an earlier chapter in the context of a PDV. Let's look at how the **Word Scanner** processes our program:

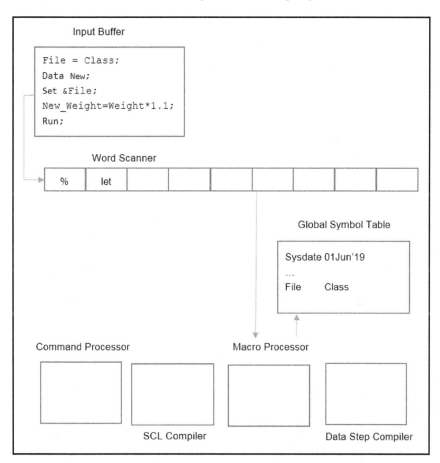

The **Macro Processor** in the preceding diagram can decipher the syntax of naming macro variables. The % symbol triggers the macro processor into action. The **Macro Processor** sends the value of the macro variable to the symbol table. The symbol table, apart from the globally declared SAS default macros, now contains a macro variable called **File** with the **Class** value.

In the following diagram, we can see that the data step compiler has been activated and it now contains the SAS commands. The data compiler stops processing the commands as soon as the ampersand (&) sign is passed on to the **Macro Processor**:

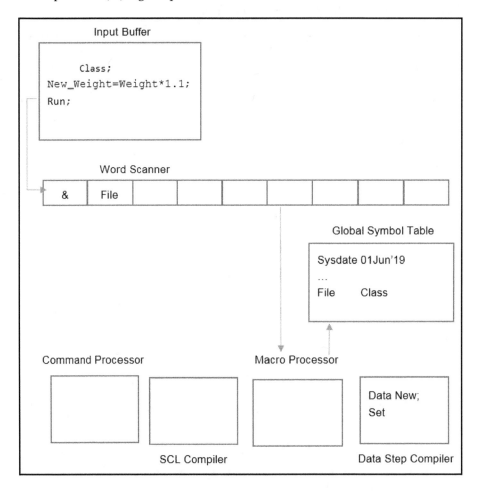

In the following diagram, we can see that the word processor finishes the processing and the data compiler has now resolved all the references to the macro variable and the Input Buffer is now empty. This completes the processing of the SAS commands that were submitted:

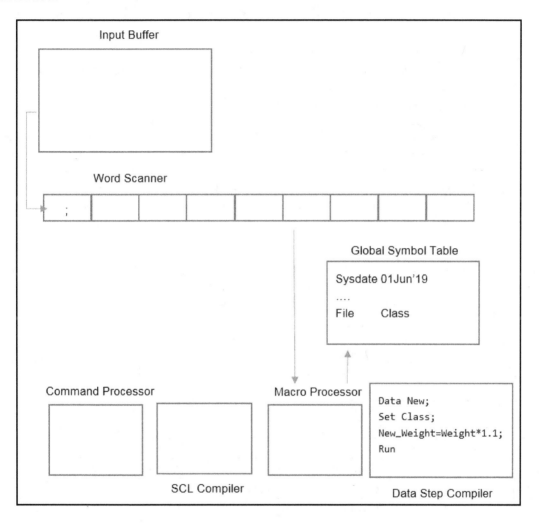

There are a few features of macros that users should be aware of:

- There is no explicit length of a macro variable
- The maximum permitted length is 32,000 characters
- The macro derives its length from the argument that is passed on as the macro variable
- Macros only store characters
- However, it is possible to evaluate the characters stored in a macro variable as numeric
- There are two types of macro variables: user-defined and automatic variables

We have already seen examples of both user-defined and automatic variables in the macro processing diagrams shown in the preceding diagrams. When a macro variable is defined outside a macro definition (in open code), the macro variable is stored as a global macro variable. When it is specified within a macro statement (beginning with %macro and ending with %mend), the variable is only available to the local SAS session and is stored as a local symbol table.

SAS already contains global macro variables. The global macro variables created by SAS (also called **automatic macro variables**) can be referenced by using the ampersand. Some of the automatic macro variables are mentioned in the following table:

Automatic macro variable	Description
&sysdate	Contains the date that an SAS job or session began executing. Format: 01Jan19
&sydate9.	Contains the date that an SAS job or session began executing. Format: 01Jan2019
&sysday	Contains the day of the week that an SAS job or session began executing
&systime	Contains the time an SAS job or session began executing
&syslast	Contains the name of the SAS data file created most recently
&syshostname	Contains the hostname of a computer
&sysncpu	Contains the current number of processors available to SAS for computations
&sysbuff	Contains text supplied as macro parameter values
&syserr	Contains a return code status set by some SAS procedures and the DATA step
&sysindex	Contains the number of macros that have started execution in the current SAS job

&sysinfo	Contains the return codes provided by some SAS procedures
&sysnobs	Contains the number of observations read from the last dataset that was closed by the previous procedure or DATA step
&sysrc	Returns a value corresponding to an error condition
&sysuserid	Contains the user ID of the current SAS process
&sysver	Provides the release number of the SAS software
&sysvlong	Provides the release number and maintenance level of SAS software, in addition to the release number
&syswarningtext	Last warning message generated in SAS log
&sysdevic	Name of the current graphics device
&sysdmg	Return code reflects the action taken on a damaged dataset
&sysfilrc	Return code set by the FILENAME statement
&sysparm	Value specified by SYSPARM = system option
&syslibrc	Return code set by the LIBNAME statement
&sysjobid	Name of current batch job or user ID
&sysenv	Foreground or background indicator

Let's also see the difference in global and local macro variables by trying to utilize the macro variable created in the macro definitions. We will use the following dataset:

```
Data Class;
   Input ClassID $ Year Age Height Weight;
   Datalines;
A1234 2013 8 85 34
A2323 2013 9 81 36
B3423 2013 8 80 31
B5324 2013 9 70 35
C2342 2013 9 80 31
D3242 2013 9 85 30
A1234 2019 14 105 64
A2323 2019 15 101 66
B3423 2019 14 100 61
B5324 2019 15 90 55
C2342 2019 15 112 70
D3242 2019 14 112 70
;

   %macro demo(File=);
   Data New;
     Set WORK.&File;
     New_Weight=Weight*1.1;
   Run;
```

```
%mend;

%demo(File=Class);

Data New;
  Set &File;
  New_Weight=Weight*1.1;
Run;
```

We have constructed a macro by starting it with the `%macro` syntax and ending it with the `%mend` syntax. This syntax is crucial as it will tell the macro processor that it needs to decipher the program in the lines between them. We haven't used a `%let` syntax to define a macro variable. Instead, we have used a method to declare the macro variables value only when we invoke the macro. The macro in this instance is called `demo`. Prior to invoking the macro, we have only run the macro definitions. This process is known as compiling the macro.

We get the following LOG, when the macro is invoked and the subsequent program is run. The program is the same within and outside the macro. The only difference is that the macro variable has been declared for the macro only:

```
73 %macro demo(File=);
 74
 75 Data New;
 76 Set WORK.&File;
 77 New_Weight=Weight*1.1;
 78 Run;
 79
 80 %mend;
 81
 82 %demo(File=Class);

NOTE: There were 19 observations read from the data set SASHELP.CLASS.
NOTE: The data set WORK.NEW has 19 observations and 6 variables.
NOTE: DATA statement used (Total process time):
      real time 0.00 seconds
      cpu time 0.01 seconds
```

As you can see, the macro variable assignment was successful and the program within the macro ran as expected. However, the macro variable was not found when the same program outside the macro tried to reference it, as shown in the following code block:

```
84 Data New;
85 Set &File;
                 _
                22
                200
WARNING: Apparent symbolic reference FILE not resolved.
ERROR: File WORK.FILE.DATA does not exist.
ERROR 22-322: Syntax error, expecting one of the following: a name, a
quoted string, ;, CUROBS, END, INDSNAME, KEY, KEYRESET, KEYS,
            NOBS, OPEN, POINT, _DATA_, _LAST_, _NULL_.

ERROR 200-322: The symbol is not recognized and will be ignored.

86 New_Weight=Weight*1.1;
87 Run;

NOTE: The SAS System stopped processing this step because of errors.
WARNING: The data set WORK.NEW may be incomplete. When this step was
stopped there were 0 observations and 2 variables.
WARNING: Data set WORK.NEW was not replaced because this step was stopped.
NOTE: DATA statement used (Total process time):
        real time 0.00 seconds
        cpu time 0.00 seconds
```

This is because the macro variable was stored as a local macro variable rather than a global macro variable.

Macro resolution tracking

In the preceding example, we are able to create the New dataset when we run the macro definitions. However, the value of the macro variable isn't written to the log. This was an instance of a single macro variable. Most macros in real life may contain multiple nested macros, with many macro variables in each nested macro. In such a situation, it is helpful to know what the resolution of the macro variable is. Adding a %PUT statement can be very helpful:

```
%Let File = Class;

Data New;
   Set &File;
   New_Weight=Weight*1.1;
```

```
Run;

%PUT The resolution of macro variable File is &File;
The %PUT statement helps write the resolution of the macro variable File to
the LOG. Below is the message in the LOG.

NOTE: There were 19 observations read from the data set SASHELP.CLASS.
NOTE: The data set WORK.NEW has 19 observations and 6 variables.
NOTE: DATA statement used (Total process time):
      real time 0.00 seconds
      cpu time 0.00 seconds
79
80 %PUT The resolution of macro variable File is &File;
The resolution of macro variable File is Class
```

The `%PUT` statement does not need to be part of the **DATA** or **PROC** step. It is useful for only displaying the resolution of macro variables in LOG. Along with `%PUT`, the following options are also useful:

OPTION	PURPOSE
ALL	Helps list all macro variables
AUTOMATIC	Lists all automatic macro variables
GLOBAL	Lists all user-defined global macro variables
LOCAL	Lists all user-defined local macro variables defined within the currently executing macro
USER	Lists all user-defined macro variables

Let's also explore a few other options, such as MLOGIC, MPRINT and SYMBOLGEN.

We will invoke the demo macro that we compiled earlier and study LOG:

```
Options MLOGIC;
%demo(File=Class);

The LOG produced is below.

75 %demo(File=Class);
 MLOGIC(DEMO): Beginning execution.
 MLOGIC(DEMO): Parameter FILE has value Class

NOTE: There were 19 observations read from the data set SASHELP.CLASS.
NOTE: The data set WORK.NEW has 19 observations and 6 variables.
NOTE: DATA statement used (Total process time):
      real time 0.00 seconds
```

```
        cpu time 0.00 seconds

    MLOGIC(DEMO): Ending execution.
```

Instead of writing a %PUT statement, it may be beneficial to use the MLOGIC option when trying to resolve multiple macro variables. The MLOGIC option has written the value of the FILE parameter to LOG. The MLOGIC resolution is already prefixed with the word MLOGIC word in LOG. If MLOGIC is in effect and the macro processor encounters a macro invocation, the macro processor displays messages that identify the following:

- The beginning of macro execution
- Values of macro parameters at invocation
- Execution of each macro definition statement
- Whether each %IF condition is true or false
- The ending of macro execution

The MLOGIC option writes the macro resolution in LOG. The MPRINT option goes a bit further and substitutes the macro variable in the program with the values that get assigned to the macro variable:

```
Options MPRINT;
%demo(File=Class);

MPRINT(DEMO): Data New;
MPRINT(DEMO): Set SASHELP.Class;
MPRINT(DEMO): New_Weight=Weight*1.1;
MPRINT(DEMO): Run;

  NOTE: There were 19 observations read from the data set SASHELP.CLASS.
  NOTE: The data set WORK.NEW has 19 observations and 6 variables.
  NOTE: DATA statement used (Total process time):
        real time 0.00 seconds
        cpu time 0.00 seconds
```

As you can see, the &File syntax has been replaced by the CLASS filename after the resolution of the macro variable.

If you are looking for a separate statement about the macro variables result then MLOGIC is best suited. Otherwise, try to use the MPRINT option:

```
Options SYMBLOGEN;
%demo(File=Class);

  SYMBOLGEN: Macro variable _SASWSTEMP_ resolves to
/folders/myfolders/.sasstudio/.images/e67748fb-709a-4652-bc7f-e969ca8431d8
  SYMBOLGEN: Some characters in the above value which were subject to macro
quoting have been unquoted for printing.
  SYMBOLGEN: Macro variable GRAPHINIT resolves to
72
73 Options SYMBOLGEN;
74
75 %demo(File=Class);
SYMBOLGEN: Macro variable FILE resolves to Class

  NOTE: There were 19 observations read from the data set SASHELP.CLASS.
  NOTE: The data set WORK.NEW has 19 observations and 6 variables.
  NOTE: DATA statement used (Total process time):
        real time 0.00 seconds
        cpu time 0.00 seconds

76
77 OPTIONS NONOTES NOSTIMER NOSOURCE NOSYNTAXCHECK;
SYMBOLGEN: Macro variable GRAPHTERM resolves to
 89
```

The SYMBOLGEN system option tells you what each macro variable resolves to by writing messages to the SAS log. This option is especially useful in spotting quoting problems, where the macro variable resolves to something other than what you intended because of a special character. It is also useful for decoding macro variables specified by a double ampersand (&&).

It is, at times, handy to store the macro variable resolution to an external file. You can store the output of MPRINT in an external file. The statements needed to store it are as follows:

```
options mprint mfile;
filename mprint 'TEMPOUT';
```

Please use this with caution—your SAS version and system access limitations may disallow storing of files to your chosen location.

Macro definition processing

Having gained a bit more knowledge about macros and their debugging, let's understand the resolution process of macro definition just as we did for macro variables earlier in this chapter.

The general syntax of the macro definition is as follows:

```
%macro macroname;
set of code statements;
....
%mend macroname;
```

As we saw earlier in the example of the DEMO macro definition we don't need to specify the macro definition name in the %mend statement. We will modify our current macro definition by adding a sorting option and call this new definition DEMO_SORT. While we have not included the macro definition name with the %mend closing statement, it is always a good coding habit to do so. As a coder, when you have nested macro definitions, having the macro definition name with %mend will help you keep track of the start and end of each macro definition.

The macro definition steps are iterative, and hence we will summarize the compilation of the macro definition. The key is to understand the interface between all the components that are going to help us compile the macro definition. You will notice that we now have a local symbol table along with a global symbol table. This is a deviation from the macro variable process we discussed earlier in the chapter.

In the macro variable process, we declared the macro variable in the data steps, and hence it was assigned as a global macro. This time, the macro variable has been defined within a macro definition and hence we have assigned it to the local symbol table:

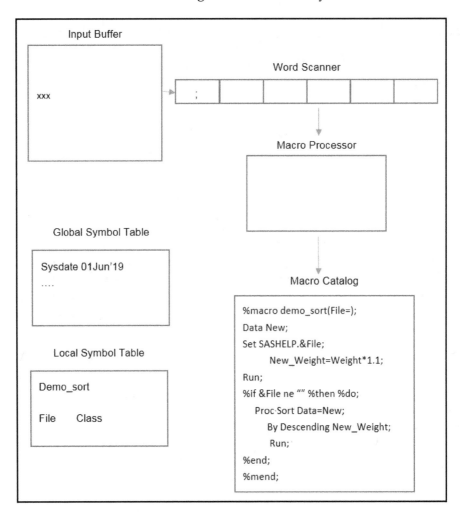

In the initial stages, the **Word Scanner** examines the input buffer and finds % followed by a non-blank character. This triggers the macro processor. The macro processor reads %demo_sort and opens the SASMACR catalog and creates a local symbol table and assigns a null value to File. The macro processor removes the token for the macro call from the input buffer and enters the File value in the local symbol table.

After this, the macro processor waits for the word scanner to tokenize the rest of the macro definition. It keeps on sending the commands to the compiler till it comes across the & sign. It then sends `&file` to the macro processor. This is an iterative process that continues between the input buffer, word scanner, macro processor, and the compiler. The key thing to note in the example earlier is that we haven't yet specified any value for the `File` macro variable defined in the macro definition. Hence, the local symbol table will not hold any value for it as yet.

As a good coding practice, always give a different name to a macro variable or macro definition in an SAS session. You could easily forget that you might have named a global macro variable with a differing value sometime earlier in your session. This might escape your attention even if debugging options are activated.

Comparing positional and keywords parameters

Till now, the chapter has used the keyword parameter to call a macro definition. In this method, we have used the parameter name with the equals sign and then specified the parameter value. This needs to be done at both the macro definition compiling and invoking stages.

However, we can also give positional parameters. In this method, the name of the parameter is supplied at the compilation stage and the value at the invoking stage. The order that you supply the values is crucial in the macro definition with the position parameters:

```
OPTIONS MLOGIC;

%macro demo_sort(File, Variable);
  Data New;
    Set WORK.&File;
    New_Weight=&Variable*1.1;
  Run;

  %if &File ne "" %then
    %do;

      Proc Sort Data=New;
        By Descending New_Weight;
      Run;
```

```
    %end;
%mend;

%demo_sort (Weight, Class);
%demo_sort (Class, Weight);

This produces an error.

MLOGIC(DEMO_SORT): Beginning execution.
MLOGIC(DEMO_SORT): Parameter FILE has value Weight
MLOGIC(DEMO_SORT): Parameter VARIABLE has value Class
ERROR: File SASHELP.WEIGHT.DATA does not exist.

NOTE: The SAS System stopped processing this step because of errors.
```

We intended to use the CLASS dataset from SASHELP in the same way we have done for the previous examples. The only change in the macro definition this time is that instead of specifying the variable name, we have defined a macro variable. Hence, we have to submit two values for the macros. We have given the correct values for the variables but we have specified them in the wrong order. The following order is correct for submitting the positional parameter macro definition:

```
%demo_sort (Class, Weight);
```

It produces the following LOG:

```
MLOGIC(DEMO_SORT): Beginning execution.
MLOGIC(DEMO_SORT): Parameter FILE has value Class
MLOGIC(DEMO_SORT): Parameter VARIABLE has value Weight

NOTE: There were 19 observations read from the data set SASHELP.CLASS.
NOTE: The data set WORK.NEW has 19 observations and 6 variables.

MLOGIC(DEMO_SORT): %IF condition &File ne "" is TRUE
```

Data-driven programming

There are various programming styles. However, data-driven programming is the most versatile as it allows you to alter your programming flow and output based on the data values. One of the key built-in macro functions that can be utilized for data-driven programming is CALL SYMPUT. It assigns a value produced in a data step to a macro variable.

SYMPUT helps create the macro variable if it doesn't already exist. It makes a macro variable assignment when it executes. Hence, be careful when trying to reference the macro variable created by CALL SYMPUT. You might experience a failure to reference the macro variable as the macro variable is only assigned during the macro execution. Hence, you cannot use the macro variable reference to retrieve the value of a macro variable in the same program in which the macro variable is being created via SYMPUT. There must be a step boundary statement to force the DATA step to execute before referencing a value in a global statement following the program. The boundary could be a RUN statement or another DATA or PROC statement.

SYMPUT puts the macro variable in the most local non-empty symbol table. A symbol table is non-empty if it contains the following:

- A value
- A computed %GOTO (a computed %GOTO contains % or & and resolves to a label)
- The &SYSPBUFF macro variable created at macro invocation time

However, there are three cases where SYMPUT creates the variable in the local symbol table, even if that symbol table is empty:

- Beginning with version 8, if SYMPUT is used after Proc SQL, the variable will be created in a local symbol table.
- If an executing macro contains a computed %GOTO statement and uses SYMPUT to create a macro variable, the variable is created in the local symbol table.
- If an executing macro uses &SYSPBUFF and SYMPUT to create a macro variable, the macro variable is created in the local symbol table.

In the `Class` dataset, we have a record of the age, height, and weight of the students in `2013` and `2019`. Let's find the tallest student in both years. The tallest student should also be the youngest in the class in the year:

```
OPTIONS MPRINT;

%macro data_driven;
  Proc Sort Data=Class Out=Sorted;
    By Descending Year Age Descending Height;
  Run;

  Data First;
    Set Sorted;
    By Descending Year Age Descending Height;

    If First.Year and First.Height then
      output First;
```

```
    Run;

    Data _NULL_;
      Set First;

      If Year eq 2019 then
        call symput("Tallest_2019", ClassID);

      If Year eq 2013 then
        call symput("Tallest_2013", ClassID);
      *Test_Call_Symput=&Tallest_2019;
      *Test to see if the macro variable can be referenced;
    Run;

    Proc Print Data=Class Noobs;
      Where Year=2019 and ClassID="&Tallest_2019";
      Title "Youngest and Tallest Child in Current Year who weighs the
    least";

    Proc Print Data=Class Noobs;
      Where Year=2013 and ClassID="&Tallest_2013";
      Title "Youngest and Tallest Child in 2013 Year who weighs the least";
    Run;

%mend data_driven;

%data_driven;
```

The preceding macro definition will fail at execution as it will produce errors due to the Test variable that we have tried to create. It references a macro variable defined by CALL SYMPUT prior to execution. On removal of the statement producing the Test variable, we get the following output:

Youngest and Tallest Child in Current Year who weighs the least				
ClassID	Year	Age	Height	Weight
D3242	2019	14	112	70

We find the youngest and tallest student in the current year who weighs the least.

Leveraging automatic global macro variables

As a programmer, it is ideal if you are aware of as many automatic global macro variables as possible. It is quite a task to remember all of them. However, there are quite a few that come in handy often. Using the automatic global variable the SYSDAY, we will try and produce two types of reports for a sales manager. If it's midweek, we will produce a detailed report. For the end of the week, our aim will be to produce a weekly summary.

We will produce a new `Sales` dataset with the date, product name, and the sales (in thousands). To try and replicate the output in the following pages, change the day condition as the following code will only produce an output if the day of the macro run is Wednesday or Friday:

```
Data Sales;
   Input SaleDate Date9. Product $ Sales;
   Format SaleDate Date9.;
   Datalines;
01Aug2019 Med1 56
02Aug2019 Med2 45
02Aug2019 Med3 48
05Aug2019 Med2 56
05Aug2019 Med3 55
06Aug2019 Med1 67
07Aug2019 NA 0
08Aug2019 Med1 54
09Aug2019 Med1 45
12Aug2019 Med2 50
13Aug2019 Med1 45
13Aug2019 Med3 53
14Aug2019 Med2 67
15Aug2019 NA 0
16Aug2019 Med2 45
;
```

We will specify debugging options to see how the data flow happens. The following macro definition will also give you a flavor of the `if...then` condition logic in a macro environment:

```
%macro dual_reporting;
Data Sales_Week;
   Set Sales;
   Week=Week(SaleDate);
Run;
```

```
   %if &sysday eq Tuesday %then
     %do;
       Title 'Mid Week Detailed Sales Report';

       Proc Print Data=Sales_Week;
         By Week;
       Run;

     %end;

   %if &sysday eq Tuesday %then
       %do;
       Title 'End of Week Sales Report Summary';

       Proc Tabulate Data=Sales_Week;
         Class Product Week;
         Var Sales;
         Table Product, Week, Sales;
       Run;

     %end;
   %mend dual_reporting;
```

If you were to run %DUAL_REPORTING on Wednesday, LOG would contain the following message:

```
SYMBOLGEN: Macro variable SYSDAY resolves to Wednesday
MLOGIC(DUAL_REPORTING): %IF condition &sysday eq Wednesday
is TRUE
MPRINT(DUAL_REPORTING): Title 'Mid Week Detailed Sales
Report';
MPRINT(DUAL_REPORTING): Proc Print Data = Sales_Week;
MPRINT(DUAL_REPORTING): By Week;
MPRINT(DUAL_REPORTING): Run;
```

The output on a Wednesday would be as follows:

Mid Week Detailed Sales Report

Week=30

Obs	SaleDate	Product	Sales
1	01AUG2019	Med1	56
2	02AUG2019	Med2	45
3	02AUG2019	Med3	48

Week=31

Obs	SaleDate	Product	Sales
4	05AUG2019	Med2	56
5	05AUG2019	Med3	55
6	06AUG2019	Med1	67
7	07AUG2019	NA	0
8	08AUG2019	Med1	54
9	09AUG2019	Med1	45

Week=32

Obs	SaleDate	Product	Sales
10	12AUG2019	Med2	50
11	13AUG2019	Med1	45
12	13AUG2019	Med3	53
13	14AUG2019	Med2	67
14	15AUG2019	NA	0
15	16AUG2019	Med2	45

For each week, we have the sales date, product, and sales. This is, in a way, a replica of the dataset except that this is presented separately for each week:

End of Week Sales Report Summary		
Product Med1		
		Sales
		Sum
Week		
30		56.00
31		166.00
32		45.00

End of Week Sales Report Summary		
Product Med2		
		Sales
		Sum
Week		
30		45.00
31		56.00
32		162.00

End of Week Sales Report Summary		
Product Med3		
		Sales
		Sum
Week		
30		48.00
31		55.00
32		53.00

The end-of-the-week report is a cross-tabulation that we have requested. For each product, we can see the week-on-week sales. The sales have been aggregated for each week and then presented for each product.

Macros that evaluate

We know that macro variables are stored as character values. The `%eval` function helps evaluate integers or logical expressions. The function converts the character data to a numeric or logical expression. After performing the evaluation, it then converts the result back to a character value and returns the value.

We will now try and evaluate some macro variables:

```
OPTIONS SYMBOLGEN;

%Let A1 = (1+0);
%Let A2 = (1+5);
%Let A3 = (10-5);
%Let A4 = (10/5);
%Let A5 = (10/3);
%Let A6 = (1-0.1);

%Let eval_A1 = %eval(&A1);
%Let eval_A2 = %eval(&A2);
%Let eval_A3 = %eval(&A3);
%Let eval_A4 = %eval(&A4);
%Let eval_A5 = %eval(&A5);
%Let eval_A6 = %eval(&A6);

%PUT eval_A1 = &eval_A1;
%PUT eval_A2 = &eval_A2;
%PUT eval_A3 = &eval_A3;
%PUT eval_A4 = &eval_A4;
%PUT eval_A5 = &eval_A5;
%PUT eval_A6 = &eval_A6;
```

In the following LOG, we get the resolution for A1, A2, A3, and A4 as we expected. For A5, the evaluation does happen, but, instead of getting a value with a decimal format, we get the whole number. A6 is not evaluated. The macro processor believes that A6 contains character strings and does not perform any mathematical or logical operation on it:

```
80 %Let eval_A1 = %eval(&A1);
SYMBOLGEN: Macro variable A1 resolves to (1+0)
81 %Let eval_A2 = %eval(&A2);
SYMBOLGEN: Macro variable A2 resolves to (1+5)
82 %Let eval_A3 = %eval(&A3);
SYMBOLGEN: Macro variable A3 resolves to (10-5)
83 %Let eval_A4 = %eval(&A4);
SYMBOLGEN: Macro variable A4 resolves to (10/5)
84 %Let eval_A5 = %eval(&A5);
SYMBOLGEN: Macro variable A5 resolves to (10/3)
```

```
85 %Let eval_A6 = %eval(&A6);
SYMBOLGEN: Macro variable A6 resolves to (1-0.1)
ERROR: A character operand was found in the %EVAL function or %IF
condition where a numeric operand is required. The condition was:
     (1-0.1)
86
87 %PUT eval_A1 = &eval_A1;
SYMBOLGEN: Macro variable EVAL_A1 resolves to 1
eval_A1 = 1
88 %PUT eval_A2 = &eval_A2;
SYMBOLGEN: Macro variable EVAL_A2 resolves to 6
eval_A2 = 6
89 %PUT eval_A3 = &eval_A3;
SYMBOLGEN: Macro variable EVAL_A3 resolves to 5
eval_A3 = 5
90 %PUT eval_A4 = &eval_A4;
SYMBOLGEN: Macro variable EVAL_A4 resolves to 2
eval_A4 = 2
91 %PUT eval_A5 = &eval_A5;
SYMBOLGEN: Macro variable EVAL_A5 resolves to 3
eval_A5 = 3
92 %PUT eval_A6 = &eval_A6;
SYMBOLGEN: Macro variable EVAL_A6 resolves to
eval_A6 =
```

To resolve the situation, we should use the `%sysevalf` macro function. It evaluates arithmetic and logical expressions using floating-point arithmetic:

```
%Let A6 = (1-0.1);

%Let eval_A6 = %sysevalf(&A6);

%PUT eval_A6 = &eval_A6;
```

The LOG showcases that the mathematical operation has been carried out and the output value is as expected:

```
73 %Let A6 = (1-0.1);
74
75 %Let eval_A6 = %sysevalf(&A6);
SYMBOLGEN: Macro variable A6 resolves to (1-0.1)
76
77 %PUT eval_A6 = &eval_A6;
SYMBOLGEN: Macro variable EVAL_A6 resolves to 0.9
eval_A6 = 0.9
```

The syntax of `%sysevalf` is as follows:

```
%sysevalf(expression, <conversion-type>)
```

In this code block, we see the following parameters:

- The `expression` parameter is an arithmetic or logical expression to evaluate.
- The `conversion-type` parameter converts the value returned by `%sysevalf` to the type of value specified. The value can then be used in other expressions that require a value of that type. The `conversion-type` parameter can be one of the following: Boolean, Ceil, Floor, and Integer.

While using the macro functions to analyze arithmetic and logical operators, we should also understand the operands and operators:

Operator	Mnemonic	Definition	Example
**		Exponentiation	5**10
+		Positive prefix	+(5+10)
-		Negative prefix	-(5+10)
¬^~	NOT	Logical not	NOT X
*		Multiplication	5*10
/		Division	5/10
+		Addition	5+10
-		Subtraction	5-10
<	LT	Less than	X<Y
<=	LE	Less than or equal to	X<=Y
=	EQ	Equal	X=Y
#	IN	Equal to one of a list	X#U V W X Y Z
¬= ^= ~=	NE	Not equal	X NE B
>	GT	Greater than	X GT B
>=	GE	Greater than or equal to	X GE B
&	AND	Logical and	X=Y and A=B
|	OR	Logical or	X=Y|A=B

Writing efficient macros

Macros are an important and powerful tool offering within SAS. But do you need to write one? What is the efficient way of writing a macro? How do you assess the benefit accrued by writing one? Does the excitement of writing a macro outweigh the benefit of writing one? These are some questions that you should ask yourself before writing macros.

For a coder, the sheer exhilaration of writing a complex macro that automates multiple tasks may be the foremost reason for writing it. However, that shouldn't be the only criterion. Do you need to write one? or Is creating a macro that tries to do a lot actually an inefficient use of the resources? Well, if the answer is that a macro is required, you may be better off writing multiple macros at times and nesting them if required into a single program file. Let's say that you want to write a macro that imports any file you want, then scans the file to produce a data quality report, imputes missing values based on various population distributions that are fitted on it, produces multiple statistical models to forecast a scenario, and then outputs the best scenario, and produces a report with all its key charts. Such a macro would be an ideal thing to have. However, writing such a macro will take up a huge amount of time and computing resources. It would be futile, considering that many of the tasks such a macro would automate are already part of the SAS environment as predefined macros or built-in wizards. Always consider the extent of the problem that your macro is trying to solve prior to writing one.

There are certain business situations where the underlying data may be changing constantly. The summary values obtained from such data may be critical inputs in the calculation of a business scenario. In such circumstances, you may want to write a macro that captures the summary values and includes them in the calculation of a business scenario. Writing a macro instead of simple code may be more efficient as it may reduce the programmer's intervention and thereby reduce the chances of human error in the calculation of the business scenario.

Assessing the benefits of writing a macro isn't a straightforward process. You will have to do a case-by-case assessment. The SAS environment you use is a compilation of hundreds of macros working in the background. Remember, you don't need to write such a complex macro environment to achieve your data or analysis goals. But, whatever your goals, use the debugging tools at hand to ensure that you are writing the most efficient code in the least amount of time possible.

Summary

In this chapter, we took our understanding of SAS and its functionalities a step forward by looking at macros. We learned how the macro processor works in conjunction with the word scanner to understand how to process macros. We looked at macro variables and writing macro definitions or programs. A few examples were presented to showcase how macros can help utilize automatic global variables, help write data-driven programs, and also evaluate arithmetic and logical operators. Toward the end of the chapter, we reviewed some macro coding best practices.

In the next chapter, we will continue to focus on macros and look at some advanced options for coding.

6
Powerful Functions, Options, and Automatic Variables Simplified

In this chapter, we will focus on learning about various functions and system options. This will build upon our earlier understanding of macro variables, definition processing, and macro resolution tracking. We will focus on the DATA step so that we can discuss macro programs and then discuss macro programming at an advanced level. We'll look at a variety of examples regarding functions and options that can make coding easier and more powerful.

The following topics will be covered in this chapter:

- Managing existing macros using:
    - NOMPRELACE and MREPLACE
    - NOMCOMPLIE and NCOMPLIE
- Understanding macro resolution using:
    - MCOMPILENOTE
    - MAUTOCOMPLOC
- Efficient coding using:
    - MACRO and NOMACRO

- Exchanging values between the DATA step and macro variables
- CALL EXECUTE
- Altering the CALL SYMPUT example
- Resolving a macro variable
- Macro quoting

NOMPREPLACE and MREPLACE

While writing macro definitions is important, protecting those we've already written can be critical in a multiple-user scenario. In this section, we will look at the roles of the NOMREPLACE and NOCOMPILE options so that we can protect our existing macro definitions.

The NOMREPLACE system option will prevent a user from overwriting a macro, even if a macro with the same name has already been compiled. Note that this will not prevent a macro variable from being rewritten. We will use the Class dataset we created in the previous chapter to highlight the role of the system option.

In the following code, the Alt dataset is being created from the Class dataset. An additional variable Dataset has been defined to take the value Alt:

```
Data Alt;
  Set Class;
  Dataset="Alt";
Run;
```

Apart from the SYMBOLGEN and MPRINT options, we have specified the NOMREPLACE option. The default system option is MREPLACE:

```
Options NOMREPLACE SYMBOLGEN MPRINT;

%Macro Value;
  %Let Target = Class;

  Data Test;
    Set &Target;
  Run;

%Mend Value;

%Value;
```

The invocation of `%Value` will be successful, and the `Test` dataset will derive its value from the `Class` dataset.

By doing this, we have compiled a new macro definition, `Replace_Value`, which overwrites the `Test` dataset that exists in the work library. The `Test` dataset now takes the value of the `Alt` dataset. The `NOMPREPLACE` value won't prevent the macro definition from compiling or the `Test` dataset from being overwritten using the new value of the `Target` macro variable:

```
%Macro Replace_Value;
  %Let Target = Alt;

  Data Test;
    Set &Target;
  Run;

%Mend Replace_Value;

%Replace_Value;
```

Let's imagine a scenario where we have passed on our existing code and macros to another user. We have created the `Replace` and `Replace_Value` macros to accommodate the difference in the value of the `Target` macro variable. If the other user decides to create a new macro definition with the names `Value` or `Replace_Value`, SAS will throw an error:

```
%Macro Replace;

%PUT "This won't run";

%Mend Replace;
%Replace;

%Macro Replace_Value;

%PUT "This won't run";

%Mend Replace_Value;
%Replace_Value;
```

Neither `Replace` nor `Replace_Value` will be created due to the `NOMPREPLACE` option being specified, which prevents the creation of macro definitions that have the same name as those stored in `WORK.SASMACR`.

The option that's specified has no effect on the stored compiled macro libraries. We get the following error on submission of the preceding code:

```
ERROR: The macro REPLACE will not be compiled because the NOMREPLACE option
is set. Source code will be discarded until a corresponding %MEND statement
is encountered.

ERROR: The macro REPLACE_VALUE will not be compiled because the NOMREPLACE
option is set. Source code will be discarded until a corresponding %MEND
statement is encountered.
```

NOMCOMPILE and NCOMPILE

The `NOCOMPILE` option goes a step further than the `NOMREPLACE` option and prevents a macro definition from being compiled. It can be used with the `NOMREPLACE` option to protect existing code from being overwritten. The `MCOMPILE` option is the default option.

The following macro definition will compile with the `Put` statement, writing the `'Macro Will Compile'` value to the LOG:

```
%Macro Compile_Test;
   %Put 'Macro Will Compile';
%Mend Compile_Test;

%Compile_Test;
```

However, with the `NOMCOMPILE` option switched on, we won't be able to compile the following macro definition, or any other macro definition in the session, for that matter, until the `MCOMPILE` option is switched off:

```
%Macro Compile_Test_Alt;
   %Put "Macro Won't Compile";
%Mend Compile_Test_Alt;

%Compile_Test_Alt;
```

This produces the following LOG:

```
ERROR: Macro compilation has been disabled by the NOMCOMPILE option. Source
code will be discarded until a corresponding %MEND statement is
encountered.
```

MCOMPILENOTE

The MCOMPILENOTE option issues a note to the LOG with details about the size and number of instructions on completion of the compilation of a macro. NOTE confirms that the compilation of the macro was completed. When the option is on and NOTE is issued, the compiled version of the macro is available for execution. A macro can successfully compile, but will still contain errors or warnings that will cause the macro to not execute as you intended.

The syntax is as follows:

```
MCOMPILENOTE=<NONE | NOAUTOCALL | ALL>
```

NONE prevents any NOTE from being written to the log.

NOAUTOCALL prevents any NOTE from being written to the log for AUTOCALL macros but does issue a NOTE to the log upon completion of the compilation of any other macro.

ALL issues an X to the log. The note contains the size and number of instructions upon the completion of the compilation of any macro.

First, we will explore the MCOMPILENOTE option with the All value:

```
OPTIONS MCOMPILENOTE=All;

%Macro Compile_Test;

   %PUT 'Macro will compile';

%Mend Compile_Test;
%Compile_Test;
```

We get the following LOG message:

```
NOTE: The macro COMPILE_TEST completed compilation without errors. 5
instructions 80 bytes.
        %Compile_Test;
NOTE: The macro COMPILE_TEST is executing from memory.
      5 instructions 80 bytes.
'Macro will compile'
```

Let's vary the code a little by changing the value of the option to `NoAutoCall` and examining the LOG:

```
OPTIONS MCOMPILENOTE=NoAutoCall;
```

Rather than calling `%Compile_Test` directly, we will compile the macro and then call it. We get the following LOG:

```
NOTE: The macro COMPILE_TEST completed compilation without errors.5
instructions 80 bytes.
```

The `NoAutoCall` option doesn't produce any significant variation in the LOG from our earlier option where ALL was specified. This option doesn't inform us about the compilation of the Auto Call macros in SAS.

The last option is where we specify the `None` option:

```
OPTIONS MCOMPILENOTE=None;
```

As expected, there will be no messages written to the LOG about the successful compilation.

In the preceding instances, the macros will not produce errors. Let's introduce an error by replacing the `%PUT` statement we used previously with a `%END` statement, without introducing a `%DO` statement:

```
OPTIONS MCOMPILENOTE=All;
%Macro Compile_Test;
%END;
%Mend Compile_Test;
%Compile_Test;
```

The ALL option sends a message to the LOG stating that the compilation has been completed with errors, as shown in the following LOG:

```
ERROR: There is no matching %DO statement for the %END. This statement will
be ignored.
 76 %Mend Compile_Test;
NOTE: The macro COMPILE_TEST completed compilation with errors.
```

The `NoAutoCall` option and the None option would have produced the same error. However, the None option won't publish a message about the success of the compilation.

NOMEXECNOTE and MEXECNOTE

NOMEXECNOTE is the default option. MEXECNOTE publishes a message about the execution of the macro. In the following macro, we have switched off a couple of other helpful options and have switched on MEXECNOTE:

```
OPTIONS MEXECNOTE NOSYMBOLGEN NOMPRINT;

%Macro Test_Log;
  Data Test;
    Set Class;
  Run;

%Mend Test_Log;

%Test_Log;
```

This writes the following message to the LOG:

```
NOTE: The macro TEST_LOG is executing from memory.
      5 instructions 80 bytes.
```

But does this mean that the note that was published confirms that the macro has executed successfully? Well, no. The system option doesn't go far enough ahead to confirm successful execution. We executed the macro in the following code block, where the reference to the Class_Alt dataset will produce an error as it does not exist in the work session. This option is less powerful than the MCOMPILENOTE option that we saw earlier. However, this option only relates to compilation and not execution:

```
%Macro Test_Log_Alt;
  Data Test;
    Set Class_Alt;
  Run;

%Mend Test_Log_Alt;

%Test_Log_Alt;
```

This still produces the same LOG Note that we received previously:

```
NOTE: The macro TEST_LOG_ALT is executing from memory.
      5 instructions 88 bytes.
ERROR: File WORK.CLASS_ALT.DATA does not exist.
```

This is in spite of the fact that an ERROR is published in the LOG.

MAUTOCOMPLOC

The `MAUTOCOMPLOC` system option can be used if you want to look at the location where the `AUTOCALL` macros are stored. It can be used as follows:

```
OPTIONS MAUTOCOMPLOC;
%PUT %LEFT( This is a);
```

The LOG contains the location of the macro:

```
73 OPTIONS MAUTOCOMPLOC;
 MAUTOLOCDISPLAY(LEFT): This macro was compiled from the autocall file
/opt/sasinside/SASHome/SASFoundation/9.4/sasautos/left.sas
 74 %PUT %LEFT( This is a);
 This is a
```

MACRO and NOMACRO

Most users will have never heard of an option that turns off macro processing in SAS. These users will even question the need for doing that. However, if you are running large processes that take a long time to complete, you will benefit from specifying the `NOMACRO` option. It prevents SAS from recognizing and processing macro language statements, macro calls, and macro variable references. In general, the item isn't recognized, and an error message is issued. If the macro facility is not used in a job, a small performance gain can be made by setting `NOMACRO` because there is no overhead when it comes to checking for macros or macro variables.

When I tried the `NOMACRO` option in SAS University Edition, I got the following error:

```
OPTIONS NOMACRO;
            _____
              11
 WARNING 11-12: SAS option MACRO is valid only at startup of the SAS System
 or startup of a SAS process. The SAS option is ignored.
```

To ensure that the `NOMACRO` option is available for your SAS environment, you may need to change the options that are saved when you set up SAS for your machine or for the client version.

Available macro functions

There are many built-in SAS macro functions available for a user. Some of them are as follows:

```
%PUT This version is &SYSVER;
%PUT Today is &SYSDAY;
%PUT My User ID is &SYSUSERID;
%PUT Current Time is &SYSTIME;
%PUT My System is &SYSSCPL;
%PUT My Operating System is &SYSSCP;
```

The preceding macro functions resolve like so:

```
73 %PUT This version is &SYSVER;
 This version is 9.4
74 %PUT Today is &SYSDAY;
 Today is Sunday
75 %PUT My User ID is &SYSUSERID;
 My User ID is sasdemo
76 %PUT Current Time is &SYSTIME;
 Current Time is 01:58
77 %PUT My System is &SYSSCPL;
 My System is Linux
78 %PUT My Operating System is &SYSSCP;
 My Operating System is LIN X64
```

You can combine the use of automatic macro variables with functions that are available in SAS. Try out this function so that you can understand the various analytical options that are available in your SAS setup:

```
%PUT License details %SYSPROD(GRAPH);
%PUT License details %SYSPROD(STAT);
%PUT License details %SYSPROD(EMINER);

73 %PUT License details %SYSPROD(GRAPH);
 License details 0
74 %PUT License details %SYSPROD(STAT);
 License details 1
75 %PUT License details %SYSPROD(EMINER);
 License details -1
```

The value of 1 denotes that the SAS product is licensed, while 0 denotes that it isn't. -1 means that the product is not available in the package or that it is not part of the software package that your organization has bought.

Exchanging values between the DATA step and macro variables

In upcoming sections, we will explore the various options that we have for exchanging values between the DATA step and macro variables and compare them.

Choosing between CALL SYMGET and CALL SYMPUT

In previous chapters, we discussed CALL SYMPUT. As you may recall, CALL SYMPUT was used to transfer DATA step variable values to a macro variable. On many occasions, you will need to transfer values from macro variables to DATA steps. Always remember the following process flow when you're trying to remember if you need to use SYMPUT or SYMGET in coding:

Without the SYMGET function, you will struggle to use macro variables in the DATA step, which includes the DATALINES and CARDS statements. In the previous chapter, we created the CLASS dataset with the ClassID, Year, Age, Height, and Weight variables. We will add a new variable, Grade, which contains the classification of the performance of various students, represented by their ClassID. The values of the performing grade will be input from macro variables that were declared before the DATA step:

```
%Let A1234=Poor;
%Let B3423=Fair;
%Let C2342=Good;
%Let D3242=Excellent;
```

We have specified the grades for four out of the six distinct ClassIDs.

Via the statement that includes the SYMGET argument, we have specified that the new Grade variable will be mapped using macro variable values:

```
Data Class;
Input ClassID $ Year Age Height Weight;
Grade = SYMGET (ClassID);
Datalines;
```

```
A1234 2013 8 85 34
A2323 2013 9 81 36
B3423 2013 8 80 31
B5324 2013 9 70 35
C2342 2013 9 80 31
D3242 2013 9 85 30
A1234 2019 14 105 64
A2323 2019 15 101 66
B3423 2019 14 100 61
B5324 2019 15 90 55
C2342 2019 15 112 70
D3242 2019 14 112 70
;
```

The macro variable values have to be mapped against the ClassID variable. Earlier, we created four macro variables with names from the ClassID variable. Then, we requested the output of the created dataset:

```
Proc Print Data=Class NoObs;
Run;
```

In the following table, we can see that the new **Grade** column has been created:

ClassID	Year	Age	Height	Weight	Grade
A1234	2013	8	85	34	Poor
A2323	2013	9	81	36	
B3423	2013	8	80	31	Fair
B5324	2013	9	70	35	
C2342	2013	9	80	31	Good
D3242	2013	9	85	30	Excellent
A1234	2019	14	105	64	Poor
A2323	2019	15	101	66	
B3423	2019	14	100	61	Fair
B5324	2019	15	90	55	
C2342	2019	15	112	70	Good
D3242	2019	14	112	70	Excellent

Since we didn't have macro variables specified for two of the six ClassIDs, we have some missing rows.

CALL EXECUTE

While discussing CALL EXECUTE, it is important to take CALL SYMPUT into consideration. As we already know, the latter is a DATA step routine that sends a character string argument to a macro variable. On the other hand, CALL EXECUTE sends a character string argument to the macro facility for immediate macro execution during the execution of the DATA step. CALL EXECUTE has been around for a long time but is still a relatively newer option as it became available in the SAS 6.07 release. The main advantage of CALL EXECUTE is that it does not require a macro or macro code, unlike CALL SYMPUT, as shown in the following code block:

```
Data Execute;
Set Class;
If Year = 2013
Then
Call Execute ('Proc Print Data = Execute; Var Age Height; Run;');
Else
Call Execute ('Proc Print Data = Execute; Run;');
Run;
```

Since we have not used any loop restrictions on the number of times `Proc Print` occurs, you will get multiple prints of the resultant data. Essentially the two types of output you will get are as follows:

Obs	Age	Height
1	8	85
2	9	81
3	8	80
4	9	70
5	9	80
6	9	85
7	14	105
8	15	101
9	14	100
10	15	90
11	15	112
12	14	112

The first `Proc Print` statement only executes if the Year condition of 2013 is met.

However, the `Proc Print` conditions don't restrict the output to 2013. This is an important point to bear in mind while trying to avoid unexpected output.

The following table is the second `Proc Print` output:

Obs	ClassID	Year	Age	Height	Weight
1	A1234	2013	8	85	34
2	A2323	2013	9	81	36
3	B3423	2013	8	80	31
4	B5324	2013	9	70	35
5	C2342	2013	9	80	31
6	D3242	2013	9	85	30
7	A1234	2019	14	105	64
8	A2323	2019	15	101	66
9	B3423	2019	14	100	61
10	B5324	2019	15	90	55
11	C2342	2019	15	112	70
12	D3242	2019	14	112	70

In the preceding table, again we haven't restricted the output to the year 2019. The `Else` statement can be a bit misleading.

If you wanted to restrict the second output to 2019 records only, try substituting the `Else` statement in the preceding code with the following:

```
Else
Call Execute ('Proc Print Data = Execute; Where Year = 2019; Run;');
```

Altering the CALL SYMPUT example

In the previous chapter, while introducing CALL SYMPUT, we looked at an example where the objective was to output records from 2019 and 2013 of the youngest people in each year with the maximum height. We had to store the ClassIDs of the people we were looking at and then pass this information on to a `Proc Print` statement to get the desired output. By using `Call Execute`, we can shorten the process to get the output in at least two steps:

```
Proc Sort Data = Class Out = Sorted;
By Descending Year Age Descending Height;
Run;
```

```
Data First;
Set Sorted;
By Descending Year Age Descending Height;
If First.Year and First.Height Then Output First;
Call Execute ('Proc Print Data = _LAST_; Where Year = 2019; Run;');
Call Execute ('Proc Print Data = _LAST_; Where Year = 2013; Run;');
Run;
```

The output will be the same as it was in previous chapter.

Resolving macro variables

We have looked at various examples of specifying and resolving macro variables throughout this book. Now, we will explore some specific instances where our current knowledge of macro variables isn't enough to overcome certain challenges.

Macro variable names within text

So far, the macro variables we have formed have had no prefix or suffix attached to them. Some macro variables, however, have had a period in front of them to segregate them from the library name, like so:

```
%Let Out = Class_;

Data &Out2013 &Out2019;
  Set Class;

  If Year EQ 2013 Then
    Output &Out2013;
  Else
    Output &Out2019;
Run;
```

In a macro variable reference, the word scanner recognizes that a macro variable name has ended when it encounters a character that is not allowed in a SAS name. To enable SAS so that it can resolve the macro variable as intended, add a period after the macro variable name:

```
%Let Out = Class_;

Data &Out.2013 &Out.2019;
  Set Class;
```

```
  If Year EQ 2013 Then
     Output &Out.2013;
  Else
     Output &Out.2019;
Run;

Proc Print Data=&Out.2013;
Run;

Proc Print Data=&Out.2019;
Run;
```

We are able to successfully create multiple datasets based on the value that's held in the `Year` variable. The following table is the dataset for the year 2013:

Obs	ClassID	Year	Age	Height	Weight
1	A1234	2013	8	85	34
2	A2323	2013	9	81	36
3	B3423	2013	8	80	31
4	B5324	2013	9	70	35
5	C2342	2013	9	80	31
6	D3242	2013	9	85	30

The following table is the dataset for 2019:

Obs	ClassID	Year	Age	Height	Weight
1	A1234	2019	14	105	64
2	A2323	2019	15	101	66
3	B3423	2019	14	100	61
4	B5324	2019	15	90	55
5	C2342	2019	15	112	70
6	D3242	2019	14	112	70

Macro variables and libraries

Macro variable resolution with library name usage is best resolved using a period.

However, you don't always need a period when the library name is derived from a macro variable:

```
OPTIONS SYMBOLGEN MPRINT;
%Let Libref = WORK;

Data &Libref.Class_Alt;
  Set Class;
Run;
```

No issues are experienced when resolving the macro variable in this way. We get the following message in the LOG:

```
75 %Let Libref = WORK;
76
SYMBOLGEN: Macro variable LIBREF resolves to WORK
77 Data &Libref.Class_Alt;
78 Set Class;
79 Run;

NOTE: There were 12 observations read from the data set WORK.CLASS.
NOTE: The data set WORK.WORKCLASS_ALT has 12 observations and 5 variables.
```

We will now use the macro variable in the SET statement and see what happens with the resolution:

```
%Let Libref = WORK;

Data Class_Alt;
  Set &Libref.Class;
Run;
```

As we can see in the following LOG, the library and filename aren't resolved:

```
73 %Let Libref = WORK;
74
75 Data Class_Alt;
SYMBOLGEN: Macro variable LIBREF resolves to WORK
76 Set &Libref.Class;
ERROR: File WORK.WORKCLASS.DATA does not exist.
77 Run;

NOTE: The SAS System stopped processing this step because of errors.
WARNING: The data set WORK.CLASS_ALT may be incomplete. When this step was
stopped there were 0 observations and 0 variables.
WARNING: Data set WORK.CLASS_ALT was not replaced because this step was
stopped.
```

To resolve this situation, we will have to place a period in the SET statement, after the macro variable:

```
Set &Libref..Class;
```

We get the following LOG message on running the entire DATA step:

```
SYMBOLGEN:  Macro variable LIBREF resolves to WORK
```

Indirect macro referencing

In our `datasets` class, the ClassIDs have multiple rows as they are present in both 2013 and 2019. Instead of having two columns, we can have both ClassID and Year in the same variable by concatenating them. ClassID can be represented by a `ClassID` macro variable, while the year value can be represented by the `Year` macro variable. To bring them together, we will have to concatenate the two macro variables. Let's say that such a variable does exist. We will try and reference the macro variable using another macro variable (this is called indirect macro referencing):

```
%Let ClassID2013 = X12342013;
*We will try and get this value in the LOG;
%Let Year = 2013;

%Macro Test;
  %Put &ClassID&Year;
%Mend;

%Test;
```

In the preceding instance, instead of getting the X12342013 value, we get the following value in the LOG:

```
SYMBOLGEN: Macro variable CLASSID resolves to A1234
SYMBOLGEN: Macro variable YEAR resolves to 2013
A12342013
```

Our output is merely the concatenation of the CLASSID and YEAR variables, rather than the indirect reference to the CLASSID2013 macro variable. We will modify the PUT statement and try and resolve the Year macro variable first:

```
%Let ClassID2013 = X12342013;
%Let Year = 2013;


%Macro Test;
  %Put &&ClassID&Year;
%Mend;
%Test;
```

The change in the order of the resolution of the macro variables will give us the intended output.

We get the following macro resolution:

```
SYMBOLGEN: && resolves to &.
SYMBOLGEN: Macro variable YEAR resolves to 2013
SYMBOLGEN: Macro variable CLASSID2013 resolves to X12342013
```

Series of macro variable references with a single macro call

Another use of an indirect reference macro is that we can generate a series of references with a single macro call by using an iterative %DO loop:

```
%Let ClassID1 = A12341;
%Let ClassID2 = A23232;
%Let ClassID3 = B34233;
%Let ClassID4 = B53244;
%Let ClassID5 = C23425;
%Let ClassID6 = D32426;

%Macro ClassIDYear;
  %Do i=1 %To 5;
    &ClassID&i
  %End;
%Mend;

%Put %ClassIDYear;
```

The output that's generated from the preceding code is a list of macro variables:

```
SYMBOLGEN: Macro variable CLASSID resolves to A1234
SYMBOLGEN: Macro variable I resolves to 1
SYMBOLGEN: Macro variable CLASSID resolves to A1234
SYMBOLGEN: Macro variable I resolves to 2
SYMBOLGEN: Macro variable CLASSID resolves to A1234
SYMBOLGEN: Macro variable I resolves to 3
SYMBOLGEN: Macro variable CLASSID resolves to A1234
SYMBOLGEN: Macro variable I resolves to 4
SYMBOLGEN: Macro variable CLASSID resolves to A1234
SYMBOLGEN: Macro variable I resolves to 5
A12341 A12342 A12343 A12344 A12345
```

This list can be used for various cross-references in macros.

Multiple ampersands

We have already seen the use of two ampersands. In SAS, there is no limit to the number of ampersands that you can place before a macro variable. How these ampersands resolve is worth understanding. The ampersands are read from left to right by the macro processor. Remember that two ampersands will always resolve to one ampersand. Let's look at some macro variable resolutions to understand multiple ampersand resolution:

```
%Let Class = A;
%Let N = 1;
%Let Class1 = B2;
%Let A1 = A1Unknown;

%PUT &Class&N;
%PUT &&Class&N;
%PUT &&&Class&N;
```

In the case of single ampersands, the `Class` macro variable is resolved with a value of `A`. The `N` macro variable is then resolved as having a value of `1`. Due to this, the combined value of the two macro variables is `A1`. In the case of `&&Class&N`, reading from left to right, the two macro ampersands are resolved to `1`.

After this, the macro processors resolve N to 1. Following this, the macro processor again reads from left to right and now the string that needs to be resolved isn't &Class but &Class1 instead. This gets resolved as B2. You can now clearly see that having an extra ampersand in front of the Class macro variable has changed the resolution:

```
78 %PUT &Class&N;
SYMBOLGEN: Macro variable CLASS resolves to A
SYMBOLGEN: Macro variable N resolves to 1
A1
79 %PUT &&Class&N;
SYMBOLGEN: && resolves to &.
SYMBOLGEN: Macro variable N resolves to 1
SYMBOLGEN: Macro variable CLASS1 resolves to B2
B2
80 %PUT &&&Class&N;
SYMBOLGEN: && resolves to &.
SYMBOLGEN: Macro variable CLASS resolves to A
SYMBOLGEN: Macro variable N resolves to 1
SYMBOLGEN: Macro variable A1 resolves to A1Unknown
A1Unknown
```

In the third instance of the triple ampersands, the first two ampersands on the left-hand side resolve to one ampersand. This makes Class resolve to A and N resolve to 1. In the end, A1 resolves to A1Unknown.

Having reviewed multiple ampersands and indirect macro variable referencing, we will look at macro quoting.

Macro quoting

On several occasions, you may have to use special characters and mnemonics as text. The user may not intend to have these treated as macro variables, but without macro quoting, the macro processor will not resolve as intended in this scenario. We will use the Proc Print code that we used to demonstrate CALL EXECUTE to highlight the use of macro quoting functions:

```
%LET Output = Proc Print; Data = Class; Where Year = 2019; Run;
```

We get the following LOG on execution:

```
75 %LET Output = Proc Print; Data = Class; Where Year = 2019; Run;
                  ____
                  180
ERROR 180-322: Statement is not valid, or it is used out of proper order.

 75 %LET Output = Proc Print; Data = Class; Where Year = 2019; Run;
                   ____
                   180
ERROR 180-322: Statement is not valid, or it is used out of proper order.
```

Even before creating the macro variable, you may have noticed that the program editor had the colors highlighted for some key words in the code. This should have alerted you that, instead of reading the code as macro variable text, the editor was treating the code as a DATA step. To mitigate this issue, we will have to use macro quoting functions. The most commonly used macro quoting functions are as follows:

- %STR and %NSTR
- %BQUOTE and %NBQUOTE
- %SUPERQ

We will rewrite the macro variable definition as follows:

```
%LET Output = %STR(Proc Print; Data = Class; Where Year = 2019; Run;);
%PUT Output is &Output;
```

We get the following LOG message in response:

```
SYMBOLGEN: Macro variable OUTPUT resolves to Proc Print; Data = Class;
Where Year = 2019; Run;
SYMBOLGEN: Some characters in the above value which were subject to macro
quoting have been unquoted for printing.
Output is Proc Print; Data = Class; Where Year = 2019; Run;
```

Macro quoting functions can be described as compilation or execution phase-related macro quoting functions. %STR and %NSTR are compilation functions. At the compilation phase, these functions help the macro processor interpret special characters as text in a macro program statement in open code or while compiling a macro. The %BQUOTE and %NRBQUOTE functions are execution functions. The resolution happens at the execution stage of a macro program statement in open code. Apart from treating these special characters as text, the macro processor resolves the rest of the macro definition normally and issues any warning messages for macro variable references or macro invocations it cannot resolve before quoting the result.

The %SUPERQ function is a unique macro quoting function as, apart from masking all special characters and mnemonic operators at the macro execution stage, it prevents further resolution of the value. Apart from this feature in %SUPERQ, the big difference with other macro quoting functions is that it only accepts the name of a macro variable as its argument. Hence, the preceding Output macro variable definition cannot be defined in the same way using the %SUPERQ function.

Using the %STR quote

Instead of applying the %STR quoting function at the start of the value of the macro variable, we can use it to quote various items in the intended macro variable value, as follows:

```
%LET Output = Proc Print%STR(;) Data = Class%STR(;) Where Year =
2019%STR(;) Run%STR(;);
%PUT Output is &Output;
```

Let's look at some further uses of %STR:

```
%Let Name = %STR(Philip%'s);
%PUT &Name;
```

In this instance, we are trying to mask the apostrophe and print the Name value as Philip's. If the macro quoting function isn't noticed, the macro variable will not execute. Apart from masking the full value with %STR, we have also placed a % sign in front of the character to be masked. This produces the following LOG:

```
SYMBOLGEN: Macro variable NAME resolves to Philip's
SYMBOLGEN: Some characters in the above value which were subject to macro
quoting have been unquoted for printing.
Philip's
```

As we mentioned previously, %STR is also helpful when dealing with mnemonic characters. It seems obvious that the resolution of this will be GE eq General Electric. However, herein lies the problem. The macro variable company resolves to GE. In the IF statement, GE = GE is not resolved. A set of nulls is resolved. Even though the answer looks correct, we aren't resolving it properly:

```
OPTIONS SYMBOLGEN MPRINT MLOGIC;

%Macro Mnemonic (Company);
  %If &Company=GE %Then
    %Put &Company eq General Electric;
  %Else
```

```
    %PUT &Company is ABC;
%Mend Mnemonic;

%Mnemonic (GE);
```

The LOG for the preceding executed macro is as follows:

```
80 %Mnemonic (GE);
 MLOGIC(MNEMONIC): Beginning execution.
 MLOGIC(MNEMONIC): Parameter COMPANY has value GE
 SYMBOLGEN: Macro variable COMPANY resolves to GE
 MLOGIC(MNEMONIC): %IF condition &Company = GE is TRUE
 MLOGIC(MNEMONIC): %PUT &Company eq General Electric
 SYMBOLGEN: Macro variable COMPANY resolves to GE
 GE eq General Electric
 MLOGIC(MNEMONIC): Ending execution.
```

Let's change the value of the macro variable to LT and view the LOG:

```
80 %Mnemonic (LT);
 MLOGIC(MNEMONIC): Beginning execution.
 MLOGIC(MNEMONIC): Parameter COMPANY has value LT
 SYMBOLGEN: Macro variable COMPANY resolves to LT
 MLOGIC(MNEMONIC): %IF condition &Company = GE is TRUE
 MLOGIC(MNEMONIC): %PUT &Company eq General Electric
 SYMBOLGEN: Macro variable COMPANY resolves to LT
 LT eq General Electric
 MLOGIC(MNEMONIC): Ending execution.
```

As we explained earlier, null values are compared in the IF statement. It doesn't matter what values are passed as the Company macro variable. The eventual result is going to be macro variable value is eq General Electric. To resolve this anomaly, we can use the %STR function.

The updated macro and LOG are as follows:

```
%Macro Mnemonic (Company);

%If &Company = %STR(GE) %Then %Put Company eq General Electric;
%Else %PUT &Company is ABC;

%Mend Mnemonic;

%Mnemonic (LT);


80 %Mnemonic (LT);
 MLOGIC(MNEMONIC): Beginning execution.
```

```
MLOGIC(MNEMONIC): Parameter COMPANY has value LT
SYMBOLGEN: Macro variable COMPANY resolves to LT
MLOGIC(MNEMONIC): %IF condition &Company = GE is FALSE
MLOGIC(MNEMONIC): %PUT &Company is ABC
SYMBOLGEN: Macro variable COMPANY resolves to LT
LT is ABC
MLOGIC(MNEMONIC): Ending execution.
```

Using the %NRSTR quote

This macro function mirrors the capabilities of `%STR` and also masks ampersands. The following macro is an example where we will get to use the `%NSRTR` quote, but first we will execute the macro without the quote and see the results:

```
%Macro IceCream;
   %PUT My favorite ice cream is Ben&Jerry;
%Mend IceCream;
%IceCream;
```

This won't resolve as expected as the macro processor has unsuccessfully tried to look for the `Jerry` macro variable. The error that's produced in the LOG states the following:

```
MLOGIC(ICECREAM): Beginning execution.
 MLOGIC(ICECREAM): %PUT My favorite ice cream is Ben&Jerry
 WARNING: Apparent symbolic reference JERRY not resolved.
 My favorite ice cream is Ben&Jerry
 MLOGIC(ICECREAM): Ending execution.
```

We will need to use `%NRSTR` in this case:

```
%Macro IceCream;
   %PUT My favorite ice cream is %NRSTR(Ben&Jerry);
%Mend IceCream;
%IceCream;
```

As you can see in the following LOG, `&Jerry` is not treated as a macro variable, and the ampersand is masked:

```
MLOGIC(ICECREAM): Beginning execution.
 MLOGIC(ICECREAM): %PUT My favorite ice cream is BenJerry
 My favorite ice cream is Ben&Jerry
 MLOGIC(ICECREAM): Ending execution.
```

Using the %BQOUTE and %NRBQOUTE quotes

`%BQUOTE` masks special characters such as parentheses, quotation marks, mathematical operators, and semi-colons. The advantage that `%BQUOTE` has over other functions, such as `%STR`, is that unmatched quotation marks, parentheses, and percent signs do not have to be used in conjunction with %. As we discussed earlier, `%BQUOTE` masks special characters when the macro is executed; this allows each quotation mark or parenthesis to be masked independently of others. `%NRBQOUTE` can also mask ampersands and the % sign. `%NRBQUOTE` differs from other quoting functions in that macro variables will be resolved where possible, but any ampersands or percent signs in the result are not viewed as operators when we use other functions such as `%EVAL` or `%PUT`.

The difference between the compilation and execution resolution of the functions can be best understood by exploring the LOG of following executed macro variables:

```
%Let Icecream = Awesome;
%Let A = %STR(&Icecream);
%Let B = %NRSTR(&Icecream);
%Let C = %BQUOTE(&Icecream);
%Let D = %NRBQUOTE(&Icecream);

%PUT A is &A;
%PUT B is &B;
%PUT C is &C;
%PUT D is &D;
```

The LOG's output is as follows:

```
A is Awesome
B is &Icecream
C is Awesome
D is Awesome
```

As we can see, the `C` and `D` macro variables don't behave as expected as both resolve at execution.

Summary

In this chapter, we have looked at various system options and macro variable functions that can help users code advanced macros. We have also explored how to use macros by correctly referencing them. Instead of simple examples, we reviewed some complex scenarios where macro quoting is necessary to ensure that the output that's generated meets requirements.

In the next chapter, we will learn about some advanced programming techniques by using Proc SQL.

Section 4: SQL in SAS

To both novice and experienced users of SQL, this part showcases how SQL offers advantages compared to the data steps discussed hitherto in this book. An understanding of SQL-generated output is developed, and the unique advantages of using macros in SQL are discussed.

This section comprises the following chapters:

- Chapter 7, *Advanced Programming Techniques Using Proc SQL*
- Chapter 8, *Deep Dive into PROC SQL*

7
Advanced Programming Techniques Using PROC SQL

SAS users can leverage SQL-based queries via Proc SQL. Both data steps and Proc SQL can generate similar outputs. However, the complexity of the code may differ. The default output generated may also differ but that can be adjusted by specifying various options available in both methods. The process of getting results in both methods differs. We will focus on the process of getting things done in Proc SQL, compare data steps and Proc SQL, and look at the specific uses of Proc SQL.

In this chapter, we will cover the following topics:

- Comparing data steps and Proc SQL
- Proc SQL joins
- Proc SQL essentials
- Dictionary tables

Comparing data steps and Proc SQL

There are some fundamental differences between how a data step and Proc SQL approach connecting tables. In a data step, we refer to this as merging tables whereas in Proc SQL we call it joining tables. The differences don't end in the naming conventions. Both aim to perform similar tasks, that is, to make connections between various tables and help bring across variables and observations from tables into a common table. The in-built functionalities available in SAS and the general differences in capabilities between data steps and Proc SQL will also yield different results or could lead to varying complexity in achieving the same output.

The following points are the fundamental differences between data steps and Proc SQL:

- The data step requires merged tables to be sorted. Proc SQL has no such requirement. This can be a big saving not just from the coding perspective but also in terms of the runtime needed to execute the sort.
- The data step requires that the merged variables have the same names. Proc SQL will not force you to rename the variables used to connect various tables.
- Data step merges cannot handle many-to-many relationships but Proc SQL can handle these.

Given the preceding three differences, it is likely that you will have come to feel that the data step is less suited to connecting tables. However, the big drawback of Proc SQL joins, when compared to the data step merge, is that the former creates a Cartesian product. A Cartesian product is a situation where all the rows and columns from the tables to be connected are joined with each other. The Cartesian product is not necessarily the output. The output will depend on the where conditions that have been given.

The following points are some further differences between data steps and Proc SQL:

- Data steps can require a few more extra steps of coding compared to Proc SQL to achieve similar results. For example, this may be necessitated by the sorting requirement of datasets while using data steps.
- When merging, duplicate matching columns across datasets are automatically overlaid. However, when joining using Proc SQL, a duplicate matching column is not automatically overlaid.
- The results are not automatically printed in a data step. A Proc print is usually required. In Proc SQL, unless you specify the NOPRINT option, the output will be printed.

As a user of SAS, you may be biased toward using Proc SQL if you have previously used SQL-based programming. The syntax is quite similar except for a few additions built into SAS while building customized SQL procedures.

Proc SQL joins

While introducing you to SAS programming we looked at the structure of the Proc SQL query. Let's look at the various join options in SAS. We will try and create a basic join without imposing any conditions.

We will use the `Class` dataset used in previous datasets and the following `Grade` dataset:

```
Data Grade;
Input ClassID $ Year Grade $;
Datalines;
A1234  2013  A
A2323  2013  A
B3423  2013  B
B5324  2013  C
C2342  2013  C
D3242  2013  D
A1234  2019  B
A2323  2019  C
B3423  2019  D
B5324  2019  B
C2342  2019  C
D3242  2019  D
;
```

The following code will produce a dataset that contains the yearly grades of the students found in the `Class` dataset:

```
Proc SQL;
        Create Table Class_Grade As
        Select *
        From Class, Grade
        ;
Quit;
```

After executing the preceding code, the key to understanding the joints doesn't lie in just exploring the results. Instead, first let's look at the messages that are produced in the LOG:

```
NOTE: The execution of this query involves performing one or more Cartesian
product joins that cannot be optimized.
WARNING: Variable ClassID already exists on file WORK.CLASS_GRADE.
WARNING: Variable Year already exists on file WORK.CLASS_GRADE.
NOTE: Table WORK.CLASS_GRADE created, with 144 rows and 6 columns.
```

The `ClassID` and `Year` variables are not replaced in the `CLASS` dataset as they are already present. The joining key hasn't been specified. Only the names of the datasets to be connected have been mentioned. As a result, we have formed what we just discussed—a Cartesian product, as shown in the following table:

Obs	ClassID	Year	Age	Height	Weight	Grade
1	A1234	2013	8	85	34	A
2	A1234	2013	8	85	34	A
3	A1234	2013	8	85	34	B
4	A1234	2013	8	85	34	C
5	A1234	2013	8	85	34	C
6	A1234	2013	8	85	34	D
7	A1234	2013	8	85	34	B
8	A1234	2013	8	85	34	C
9	A1234	2013	8	85	34	D
10	A1234	2013	8	85	34	B
11	A1234	2013	8	85	34	C
12	A1234	2013	8	85	34	D
13	A1234	2019	14	105	64	A
14	A1234	2019	14	105	64	A
15	A1234	2019	14	105	64	B

We had 12 rows each in the connected datasets. Hence, the connected dataset has 12x12 = 144 rows. The Cartesian product cannot be optimized as SAS has no option other than to produce a 12x12 matrix in this instance.

This was an instance of the default join. In the data step, the default is the one-to-one outer join. We will learn more about this type of join in this chapter.

We will evaluate the following joins in this chapter:

- Inner join
- Left join
- Right join
- Full join
- One-to-many join
- Many-to-many join

Inner join

In an inner join, the output contains data from connecting tables for the matching records only. As you can see in the following inner-join Venn diagram, only the shaded portion of the intersection of the two datasets will be represented in the output dataset. The matching key can be specified on one or multiple conditions:

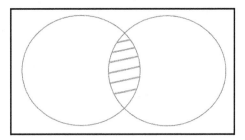

A hospital may be interested in matching the records of those requiring a transplant and potential donors identified. They might want a list that contains information about the donor and recipient in a single dataset. The list of donors without a matching recipient may be of no interest to the hospital. In such a case, an inner join query will be the best choice to create the dataset.

I have created a dataset called `Interests` that can be matched to the `Class` dataset at a `ClassID` level. The data in `Interests` isn't specific to the `Grade` of the student:

```
Data Interests;
Input ClassID $ Music Sports Drama Photography;
Datalines;
A1234 1 1 1 0
A2323 1 0 1 .
B3423 1 1 1 0
D3242 . 0 1 1
E4234 1 1 . 1
F5642 1 1 1 0
G6534 0 1 1 .
D4234 1 . 0 1
S3576 1 0 0 1
;
```

We will now attempt an inner join to the `Class` table:

```
Proc Sql;
        Create table Inner_Join as
        Select *
        From Class
```

```
                    Inner Join Interests
                    On Class.ClassID=Interests.ClassID
                    ;
Quit;
```

The following LOG messages are produced:

```
WARNING: Variable ClassID already exists on file WORK.INNER_JOIN.
NOTE: Table WORK.INNER_JOIN created, with 8 rows and 9 columns.
```

The output dataset created is as follows:

Obs	ClassID	Year	Age	Height	Weight	Music	Sports	Drama	Photography
1	A1234	2013	8	85	34	1	1	1	0
2	A2323	2013	9	81	36	1	0	1	.
3	B3423	2013	8	80	31	1	1	1	0
4	D3242	2013	9	85	30	.	0	1	1
5	A1234	2019	14	105	64	1	1	1	0
6	A2323	2019	15	101	66	1	0	1	.
7	B3423	2019	14	100	61	1	1	1	0
8	D3242	2019	14	112	70	.	0	1	1

As we can see, Class has 12 rows and 6 distinct ClassIDs. The Interests dataset has nine rows and nine distinct ClassID. The intersection of the two datasets leads to a table with only four distinct ClassIDs. We end up with eight rows in the output dataset as each of the four matched ClassIDs has a row each for 2013 and 2019. All columns from the Class dataset have been retained. The only column that theoretically wasn't included from the Interests table is the ClassID column. This, however, has no bearing as the variable already exists in the Class dataset.

Left join

The dataset on the left-hand side of a join will be present in the output dataset irrespective of the matching records that the connecting dataset has. The following diagram is an illustration of a left join Venn diagram:

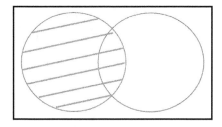

In the previous hospital inner join example, we had a list where recipients and donors matched. The hospital might be interested in retaining its recipient list and flagging instances when the donor is available. In such an instance a left joint becomes important.

In the case of the Class table, if we want to retain all the information and add on the information of the Interests table, we can use a left joint where only the matching records from the latter table will be added on. All records of the Class table will be retained:

```
Proc Sql;
        Create table Left_Join as
        Select A., B.
        From Class as A
        Left Join Interests as B
        On A.ClassID=B.ClassID
        ;
Quit;
```

This will result in the following output:

Obs	ClassID	Year	Age	Height	Weight	Music	Sports	Drama	Photography
1	A1234	2013	8	85	34	1	1	1	0
2	A1234	2019	14	105	64	1	1	1	0
3	A2323	2013	9	81	36	1	0	1	.
4	A2323	2019	15	101	66	1	0	1	.
5	B3423	2013	8	80	31	1	1	1	0
6	B3423	2019	14	100	61	1	1	1	0
7	B5324	2019	15	90	55	.	.	.	.
8	B5324	2013	9	70	35	.	.	.	.
9	C2342	2013	9	80	31	.	.	.	.
10	C2342	2019	15	112	70	.	.	.	.
11	D3242	2013	9	85	30	.	0	1	1
12	D3242	2019	14	112	70	.	0	1	1

As we can see from the output, instances when the `ClassID` in the left table isn't present in the right table, all the observations of the variables from the right table have been set to missing.

Note that the **Photography** variable has missing values in observations 3 and 4 even though the `ClassID` has matched to the `Interests` table. We know that the `ClassID` matched as the rest of the variables from `Interests` have a populated value. As a data user, you need to be careful about interpreting data. We will run a left join query where the only variable of interest from the second table is the Photography variable:

```
Proc Sql;
        Create table Left_Join_Inference as
        Select A.*, B.Photography
        From Class as A
        Left Join Interests as B
        On A.ClassID=B.ClassID
        ;
Quit;
```

In the preceding two outputs, the values of the Photography variable are the same for the `ClassID`. What has changed is the interpretation. Our dilemma lies between the following two choices:

- The A2323, B5324, and C2342 ClassIDs are not present in the `Interests` dataset and hence the variable added has no values.
- Some or all the missing values of the variable are absent because `Interests` doesn't have the values populated for the variable.

This is a dilemma that the user will face when using a left join. There are ways to overcome this dilemma by creating subqueries or variables notify us to inform whether the missing value is a source data issue or a non-matching join variable issue. In any circumstance, be careful to interpret the result of the left join for missing values in the dataset on the right-hand side of the join. The following table is a left join output dataset with **Photography** added:

Obs	ClassID	Year	Age	Height	Weight	Photography
1	A1234	2013	8	85	34	0
2	A1234	2019	14	105	64	0
3	A2323	2013	9	81	36	.
4	A2323	2019	15	101	66	.
5	B3423	2013	8	80	31	0
6	B3423	2019	14	100	61	0
7	B5324	2019	15	90	55	.
8	B5324	2013	9	70	35	.
9	C2342	2013	9	80	31	.
10	C2342	2019	15	112	70	.
11	D3242	2013	9	85	30	1
12	D3242	2019	14	112	70	1

The Venn diagram representation for a left join in can be quite confusing when taken at face value. It seems to suggest that there is a small area of intersection in connecting datasets while using a left join. What would happen if the `Interests` dataset suffered from a data quality issue where the prefix 9 got added on to the ClassID?:

```
Data Interests;
Set Interests;
ClassID=Compress("9"||ClassID);
Run;
```

Our output, in this case, would just contain all the records from the dataset `ClassID`:

Obs	ClassID	Year	Age	Height	Weight	Music	Sports	Drama	Photography
1	A1234	2013	8	85	34	.	.	.	.
2	A1234	2019	14	105	64	.	.	.	.
3	A2323	2013	9	81	36	.	.	.	.
4	A2323	2019	15	101	66	.	.	.	.
5	B3423	2013	8	80	31	.	.	.	.
6	B3423	2019	14	100	61	.	.	.	.
7	B5324	2019	15	90	55	.	.	.	.
8	B5324	2013	9	70	35	.	.	.	.
9	C2342	2013	9	80	31	.	.	.	.
10	C2342	2019	15	112	70	.	.	.	.
11	D3242	2013	9	85	30	.	.	.	.
12	D3242	2019	14	112	70	.	.	.	.

No records matched from `Interests`. However, we did get the variable names populated from `Interests`. This is due to the PDV that was formed as a result of the left join.

Right join

The only difference from the left join is that the records from the right table are selected whether a match is found or not. In Proc SQL the order of the tables in the join is critical. However, this isn't the case in the merge process using data steps. The following diagram shows a right join Venn diagram:

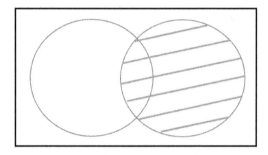

We will now keep the `Interests` table as the right-hand table and attempt a right join with the `Class` table:

```
Proc Sql;
            Create table Right_Join as
            Select A., B.
            From Class as A
            Right Join Interests as B
            On A.ClassID=B.ClassID
            ;
Quit;
```

As you can see, the code is similar to what we used earlier except for the fact that we have now specified a right join instead of an inner, or left join. This does produce some unintended results:

Obs	ClassID	Year	Age	Height	Weight	Music	Sports	Drama	Photography
1	A1234	2013	8	85	34	1	1	1	0
2	A1234	2019	14	105	64	1	1	1	0
3	A2323	2013	9	81	36	1	0	1	.
4	A2323	2019	15	101	66	1	0	1	.
5	B3423	2013	8	80	31	1	1	1	0
6	B3423	2019	14	100	61	1	1	1	0
7	D3242	2013	9	85	30	.	0	1	1
8	D3242	2019	14	112	70	.	0	1	1
9		.	.	.	.	1	.	0	1
10		.	.	.	.	1	1	.	1
11		.	.	.	.	1	1	1	0
12		.	.	.	.	0	1	1	.
13		.	.	.	.	1	0	0	1

We weren't expecting the `Year`, `Age`, `Height`, and `Weight` variables from the `Class` table to have missing values where the `ClassID` doesn't match. However, due to the `ClassID` being present on both tables, the values of the `ClassID` from the left table have superseded the values from the right table even though a right join has been requested. Certainly, if we are doing a right join, we will want to retain the records of the right table. We don't want to end up finding our `ClassID` from the right table is set to missing for 9 to 13 observations even though the variables from the contributing table have values for these observations.

The following LOG message best summarizes the reason for our problem:

```
WARNING: Variable ClassID already exists on file WORK.RIGHT_JOIN.
NOTE: Table WORK.RIGHT_JOIN created, with 13 rows and 9 columns.
```

To mitigate this we will have to stop relying on the `select *` option and specify the variables that are needed in the connected dataset:

```
Proc Sql;
        Create table Right_Join as
        Select B.*, A.Year, Age, Height, Weight
        From Class as A
        Right Join Interests as B
        On A.ClassID=B.ClassID
        ;
Quit;
```

Note that the order of the variables in the `Select` statement has no bearing on the join condition. We specified the variables from the right-hand dataset first in the Select statement even though we requested a right-hand join. We get the following table as the output:

Obs	ClassID	Music	Sports	Drama	Photography	Year	Age	Height	Weight
1	A1234	1	1	1	0	2013	8	85	34
2	A1234	1	1	1	0	2019	14	105	64
3	A2323	1	0	1	.	2013	9	81	36
4	A2323	1	0	1	.	2019	15	101	66
5	B3423	1	1	1	0	2013	8	80	31
6	B3423	1	1	1	0	2019	14	100	61
7	D3242	.	0	1	1	2013	9	85	30
8	D3242	.	0	1	1	2019	14	112	70
9	D4234	1	.	0	1	.	.	.	.
10	E4234	1	1	.	1	.	.	.	.
11	F5642	1	1	1	0	.	.	.	.
12	G6534	0	1	1	.	.	.	.	.
13	S3576	1	0	0	1	.	.	.	.

As a result of the modification where we have deselected the `ClassID` variable from the left-hand table, we get the required output with no `ClassID` set to missing.

Full join

The first thing to remember about a full join is that it is not a cross join or what we might call a default Cartesian product. A full join does not have a multiplier effect. It doesn't match each record to every record in the connecting table. So what is a full join? A full join can be considered a combination of an inner join with a left join plus right join where the records are only represented once. As a SAS learner, try and ensure that you don't get confused by the concepts of a full and a cross join. The following diagram is a full join Venn diagram:

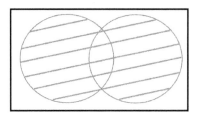

We will use the `Coalesce` function to ensure that we have no missing `ClassID`. We could have used the `Coalesce` function in the discussion we had about right joins where the `ClassID` from the right-hand side table was set to missing:

```
Proc Sql;
        Create table Full_Join as
        Select Coalesce(A.ClassID, B.ClassID), A.Year, A.Age,
A.Height, A.Weight, B.Music, B.Sports, B.Drama,
B.Photography
        From Class as A
        Full Join Interests as B
        On A.ClassID=B.ClassID
        ;
Quit;
```

The resultant dataset does not have 12x9 = 108 rows. 108 rows will only be produced in the case of a cross join. We have 12 rows in the `Class` table and 9 in the `Interests` table. However, the output has only 17 rows:

Obs	_TEMA001	Year	Age	Height	Weight	Music	Sports	Drama	Photography
1	A1234	2013	8	85	34	1	1	1	0
2	A1234	2019	14	105	64	1	1	1	0
3	A2323	2013	9	81	36	1	0	1	.
4	A2323	2019	15	101	66	1	0	1	.
5	B3423	2013	8	80	31	1	1	1	0
6	B3423	2019	14	100	61	1	1	1	0
7	B5324	2019	15	90	55	.	.	.	.
8	B5324	2013	9	70	35	.	.	.	.
9	C2342	2013	9	80	31	.	.	.	.
10	C2342	2019	15	112	70	.	.	.	.
11	D3242	2013	9	85	30	.	0	1	1
12	D3242	2019	14	112	70	.	0	1	1
13	D4234	.	.	.	.	1	.	0	1
14	E4234	.	.	.	.	1	1	.	1
15	F5642	.	.	.	.	1	1	1	0
16	G6534	.	.	.	.	0	1	1	.
17	S3576	.	.	.	.	1	0	0	1

It's difficult to say which of the joins is the most popular. It all depends on the task at hand or the general nature of the database you 're working with.

One-to-many join

A few readers may have been able to spot instances of one-to-many joins in the aforementioned examples in this chapter. Let me change the datasets that we use to discuss this join. After going through the example, I would encourage readers to look at the inner, left, right, and full join examples and see if they can spot instances of one-to-many joins.

Let's assume we have datasets X and Y:

```
Data X;
Input ID VarTabA VarTabB;
Datalines;
1 66 77
2 55 66
3 77 55
;
```

```
Data Y;
Input ID Category $ VarTabC VarTabD;
Datalines;
1 A 60 70
1 B 50 60
2 A 50 60
3 C 70 50
;
```

The datasets formed with X and Y have the same ID variables. However, the Y dataset has an instance where the ID is repeated and has two different sets of values of VarTabC and VarTabD. The following tables shows a one-to-many dataset:

Obs	ID	VarTabA	VarTabB
1	1	66	77
2	2	55	66
3	3	77	55

Obs	ID	Category	VarTabC	VarTabD
1	1	A	60	70
2	1	B	50	60
3	2	A	50	60
4	3	C	70	50

The following query was run to get a one-to-many match:

```
Proc Sql;
   Create table One_to_One as
Select Coalesce(A.ID, B.ID) as ID, VarTabA, VarTabB, VarTabC,
VarTabD
   From X as A, Y as B
```

```
   Where A.ID=B.ID
   ;
Quit;
```

Due to the data structure, the resulting join leads to a one-to-many match where one record in table X is mapped to multiple records in table Y where the ID value equals 1:

ID	VarTabA	VarTabB	VarTabC	VarTabD
1	66	77	60	70
1	66	77	50	60
2	55	66	50	60
3	77	55	70	50

The preceding query we wrote is an inner join. Don't get confused with the change in the syntax. The query is the same as the following:

```
Proc Sql;
   Create table One_to_Many as
Select Coalesce(A.ID, B.ID) as ID, VarTabA, VarTabB, VarTabC,
VarTabD
   From X as A Inner Join Y as B
   On A.ID=B.ID
   ;
Quit;
```

Many-to-many join

For a many-to-many join, we use the Y dataset and create a new dataset Z:

```
Data Z;
Input ID Category $ VarTabE VarTabF;
Datalines;
1 A 10 70
1 B 20 60
2 A 30 40
2 D 40 50
3 C 70 50
;
```

The dataset created is:

ID	Category	VarTabE	VarTabF
1	A	10	70
1	B	20	60
2	A	30	40
2	D	40	50
3	C	70	50

We will use the same code as above but just change the file names to Y and Z. We get the following output:

ID	VarTabE	VarTabF	VarTabC	VarTabD
1	10	70	60	70
1	10	70	50	60
1	20	60	60	70
1	20	60	50	60
2	30	40	50	60
2	40	50	50	60
3	70	50	70	50

For ID 1, we get four rows of output. There were two ID rows each in both the datasets. So essentially for ID 1, what we have is a Cartesian product of the ID restricted to the particular ID.

Proc SQL essentials

Having looked at join queries, we have already got a feel for how Proc SQL works. Let us review the essentials to put us on a firm footing before we explore macros in Proc SQL.

Subsetting

The simplest way to subset a dataset is by using the `Where` statement in Proc SQL. The general syntax of such a query is:

```
Proc SQL;
            Select *
            From
            Where
            ;
    Quit;
```

You can also subset using the drop and keep option. This is a good way to restrict the variables that are needed in the output dataset. Remember that subsetting is crucial in Proc SQL as the default is the Cartesian product. Any amount of filtering that can be done will reduce the number of rows processed and thereby make the querying process faster.

Other options include first applying the where condition which would reduce the largest number of records from the output dataset. In a large dataset, the order of the where condition can lead to a significant saving in processing times.

One of the reasons we explored joins before the essentials is that we can use joins to subset our dataset.

We can use the example of a bank where the managers are interested in knowing about joint account holders who have opened the same product but in a different sequence of joint ownership, for instance, where Miss A and Mr B have opened a product C but in one instance Miss A is named first and in another instance Mr B is named first. The managers want to find such instances to understand if this is a case of duplicate accounts being opened or if the joint account holders required separate first and second name account combinations:

```
Data Products;
Input Customer1 $ Customer2 $ Product $12.;
Datalines;
RT0001 RT1101 CreditCard
RT1101 RT0001 CreditCard
RT1401 RT1200 Saving
RT1002 RT1405 Current
RQ1300 RO1400 Mortgage
RO1400 RQ1300 Mortgage
RX4599 RM1001 CurrentExtra
RM1001 RX4599 Current
;
```

The following dataset is produced:

Obs	Customer1	Customer2	Product
1	RT0001	RT1101	CreditCard
2	RT1101	RT0001	CreditCard
3	RT1401	RT1200	Saving
4	RT1002	RT1405	Current
5	RQ1300	RO1400	Mortgage
6	RO1400	RQ1300	Mortgage
7	RX4599	RM1001	CurrentExtra
8	RM1001	RX4599	Current

To subset using a join, we will join the `Products` dataset with itself using the following query:

```
Proc Sql;
        Create Table Duplicate_Products as
        Select A.*,
        From Products as A
        Inner Join Products as B
        On A.Customer1=B.Customer2
        And A.Customer2=B.Customer1
        And A.Product=B.Product
    ;
Quit;
```

We have used a triple join in this instance. So far in this chapter we have used only a single join condition. The triple join ensures that we are able to identify the combination of customer accounts where the product type is common. We get the following output:

Obs	Customer1	Customer2	Product
1	RT0001	RT1101	CreditCard
2	RT1101	RT0001	CreditCard
3	RQ1300	RO1400	Mortgage
4	RO1400	RQ1300	Mortgage

To do the same using a data step for merging, we would have to sort the dataset as we discussed earlier. Also rather than merging to the same dataset we will have to create another dataset which is a replica of `Products`, and then execute the merge:

```
Proc Sort Data = Products;
By Customer1 Customer2 Product;
Run;

Data Products_Alt (Rename = (Customer1=Customer2 Customer2=Customer1));
Set Products;
Run;

Proc Sort Data = Products_Alt;
        By Customer1 Customer2 Product;
Run;

Data Duplicate_Products_Alt;
        Merge Products (in=a) Products_Alt (in=b);
        By Customer1 Customer2 Product;
        If a and b;
Run;
```

The output dataset, `Duplicate_Products_Alt` is the same as `Duplicate_Products` as shown in the previous output.

Grouping and summarizing

The following is a list of summary functions that can be requested in SAS via Proc SQL:

Summary function	Description
AVG, MEAN	Mean or average of values
COUNT, FREQ, N	Number of non-missing values
CSS	Corrected sum of squares
CV	Coefficient of variation (percent)
MAX	Largest value
MIN	Smallest value
NMISS	Number of missing values
PRT	Probability of a greater absolute value of student's t
RANGE	Range of values
STD	Standard deviation
STDERR	Standard error of the mean

SUM	Sum of values
T	Student's t value for testing a hypothesis
USS	Uncorrected sum of squares
VAR	Variance

Summary function results can be stored as a dataset or printed to the output destination:

```
Proc SQL;
Select Avg(Age) as Avg_Age, Nmiss(Age) as Missing_Age,
Std(Age) as  Std_Age
          From Class
             ;
Quit;
```

This produces the three requested sets of measures:

Avg_Age	Missing_Age	Std_Age
11.58333	0	3.088346

In the `Class` dataset, we have the `Year` variable . The metrics are produced irrespective of the Year. If we want to have the metric by `Year`, we can request grouping at a by-group level:

```
Proc SQL;
Select Avg(Age) as Avg_Age, Nmiss(Age) as Missing_Age,
Std(Age) as  Std_Age
          From Class
          Group by Year
             ;
Quit;
```

The output will now contain two rows—one each for 2013 and 2019:

Avg_Age	Missing_Age	Std_Age
8.666667	0	0.516398
14.5	0	0.547723

There is an aspect of summarizing that causes confusion among users. The confusion is caused by the `Having` clause. Users get confused between when to use the `Group by` and `Having` clauses. Some users also may also get confused about the role of the `Where` and `Having` clauses.

To keep it simple, remember that a `Where` clause comes before a `Group by` and the `Having` clause follows the `Group by`. None of the three are essential to running a Proc SQL query.

We will use all three clauses on the `Class` dataset. Let's restrict the `Class` dataset to instances where the weight is greater than 30. From the resultant records, let's select students whose height is more than the average height of the students for each `Year`:

```
Proc SQL;
            Create table GT_Avg_2019 as
                        Select *, Avg(Height) as Avg_Height
            From Class
            Where Weight GT 30
            Group by Year
            Having Height GT Avg_Height
            ;
Quit;
```

We get the following output:

Obs	ClassID	Year	Age	Height	Weight	Avg_Height
1	B3423	2013	8	80	31	79.200
2	C2342	2013	9	80	31	79.200
3	A2323	2013	9	81	36	79.200
4	A1234	2013	8	85	34	79.200
5	A1234	2019	14	105	64	103.333
6	D3242	2019	14	112	70	103.333
7	C2342	2019	15	112	70	103.333

As you can see, we have created a new variable called `Avg_Height`. `ClassID`s with weight LE 30 are not part of the output dataset. The average height has been calculated at the `Year` level and records in the output dataset have a height greater than or equal to the average height.

Dictionary tables

Dictionary tables are special read-only Proc SQL tables or views. They retrieve information about all the SAS libraries, SAS datasets, SAS system options, and external files that are associated with the current SAS session. For example, the `DICTIONARY.COLUMNS` table contains information such as `name`, `type`, `length`, and `format`, about all columns in all tables that are known to the current SAS session.

Proc SQL automatically assigns the DICTIONARY libref. To get information from DICTIONARY tables, specify DICTIONARY.table-name in the FROM clause in a SELECT statement in PROC SQL.

A list of dictionary tables and associated SASHELP views is as follows:

DICTIONARY Table	SASHELP View	Description
CATALOGS	VCATALG	Contains information about known SAS catalogs.
CHECK_CONSTRAINTS	VCHKCON	Contains information about known check constraints.
COLUMNS	VCOLUMN	Contains information about columns in all known tables.
CONSTRAINT_COLUMN_USAGE	VCNCOLU	Contains information about columns that are referred to by integrity constraints.
CONSTRAINT_TABLE_USAGE	VCNTABU	Contains information about tables that have integrity constraints defined on them.
DATAITEMS	VDATAIT	Contains information about known information map data items.
DESTINATIONS	VDEST	Contains information about known ODS destinations.
DICTIONARIES	VDCTNRY	Contains information about all DICTIONARY tables.
ENGINES	VENGINE	Contains information about SAS engines.
EXTFILES	VEXTFL	Contains information about known external files.
FILTERS	VFILTER	Contains information about known information map filters.
FORMATS	VFORMAT VCFORMAT	Contains information about currently accessible formats and informats.
FUNCTIONS	VFUNC	Contains information about currently accessible functions.
GOPTIONS	VGOPT VALLOPT	Contains information about currently defined graphics options (SAS/GRAPH software). SASHELP.VALLOPT includes SAS system options as well as graphics options.

INDEXES	VINDEX	Contains information about known indexes.
INFOMAPS	VINFOMP	Contains information about known information maps.
LIBNAMES	VLIBNAM	Contains information about currently defined SAS libraries.
MACROS	VMACRO	Contains information about currently defined macro variables.
MEMBERS	VMEMBER VSACCES VSCATLG VSLIB VSTABLE VSTABVW VSVIEW	Contains information about all objects that are in currently defined SAS libraries. SASHELP.VMEMBER contains information for all member types; the other SASHELP views are specific to particular member types (such as tables or views).
OPTIONS	VOPTION VALLOPT	Contains information about SAS system options. SASHELP.VALLOPT includes graphics options as well as SAS system options.
REFERENTIAL_CONSTRAINTS	VREFCON	Contains information about referential constraints.
REMEMBER	VREMEMB	Contains information about known remembers.
STYLES	VSTYLE	Contains information about known ODS styles.
TABLE_CONSTRAINTS	VTABCON	Contains information about integrity constraints in all known tables.
TABLES	VTABLE	Contains information about known tables.
TITLES	VTITLE	Contains information about currently defined titles and footnotes.
VIEWS	VVIEW	Contains information about known data views.
VIEW_SOURCES	Not available	Contains a list of tables (or other views) referenced by the SQL or DATASTEP view, and a count of the number of references.

Dictionary tables and columns are powerful tools that can help you write advanced queries. The following queries for accessing dictionary tables and columns can be written in the working SAS session of our chapter:

```
Proc Sql;
            Select * From Dictionary.Tables;
            Select * From Dictionary.Columns
                    Where Name = 'Class';
    Quit;
```

Run these queries and explore the results.

Summary

In this chapter, we learned about connecting data steps using Proc SQL instead of using data steps. We explored the various types of join that help connect datasets using Proc SQL. Having reviewed the pros and cons of connecting datasets in Proc SQL and data steps, we found that sorting is essential in the latter method of connecting datasets. This may mean that data step merging could be a good alternative for smaller datasets but it may lead to processing delays on a large dataset due to the sorting requirement.

We also reviewed how we can create data subsets and summarize data. We used an example where the WHERE, GROUP BY and HAVING clauses were used together to highlight the role of each of these clauses. In previous chapters, we touched upon the concept of Dictionary tables and Columns. In this chapter, we looked at an exhaustive list of options available to leverage this feature using Proc SQL.

In the next chapter, we will delve into Proc SQL by looking at data manipulation techniques, creating indexes and views, and using macros.

Deep Dive into PROC SQL

8

Proc SQL is not just about learning how Cartesian products are formed and how tables are joined. In this chapter, we will go beyond these concepts and learn about SAS views. SAS views is a concept that we will be mentioning for the first time in this book. Manipulating data and macros has already been discussed in this book at length. Now, we will look at the differences in coding in Proc SQL and how some output can be generated faster than it can be by using DATA steps. We will also see how powerful and unique some of the Macro options are in Proc SQL.

In this chapter, we will explore the following topics:

- SAS views in Proc SQL
- Manipulating data in Proc SQL
- Identifying duplicates using Proc SQL
- Creating an index in Proc SQL
- Macros in Proc SQL

SAS views in Proc SQL

In a risk regulatory framework, an analyst will be required to build models to assess credit, market, operational, or non-financial risk and submit the models to the regulator for approval. In such a scenario, the analyst will have to store the datasets that are so that regulators can recreate the results independently if needed on a static dataset. If the same analyst worked in the pharmaceutical clinical trial sector, there would be a similar need for preserving the datasets that were used.

But not all analysis needs to be scrutinized in a similar way or needs to be rerun on static databases. In a large car dealership, there might be a surge in orders that are booked at the end of the month, during a promotion, or before the financial year ends. Management might be interested in a report of daily sales figures. In this instance, the query would remain the same but it might be run at the end of each day or multiple times during the day. A SAS analyst can create a Data or SQL view in this instance. The SAS view is a SAS dataset with a different extension that will fetch data values from the underlying files.

The SAS view will save disk space compared to a dataset. The view will store instructions for where to find the data. Apart from this, the view will also contain descriptor information, which will contain the data type and lengths of the variables. But the biggest advantage in the case of the car dealership scenario will be that by running the SAS view, the analyst can ensure that the input dataset will always be current since the data is extracted by SAS views at execution time.

SAS views have more advantages than just this. Some of them are as follows:

- It prevents analysts from having to recreate the query and therefore having to redo data processing tasks or get different results from peers.
- Sensitive information can also be shielded as the SAS view will be predefined and the analyst doesn't need to explore the underlying tables.
- The complexity of data processing doesn't need to be showcased to everyone. SAS views will present streamlined information to the analyst without showcasing the complexity.

Viewers can be created using SAS DATA steps and Proc SQL. Unless the user wants to use DO LOOPS or IF-THEN processing, which is only available in the DATA step views, Proc SQL views should be preferred. Unlike DATA steps, the Proc SQL view can update the underlying data that's being used by the view. SQL can be used to subset the data that needs to be used, whereas the data to be used in the data view cannot be qualified. Hence, the entire DATA step view needs to be loaded into memory before part of it can be discarded. More types of WHERE clauses are supported by Proc SQL and it has a CONNECT component that allows us to send SQL statements to a **database management system** (**DBMS**) by using the pass-through facility.

In this chapter, we will only be discussing the native view, and not the interface view. The former is created by the DATA step or Proc SQL, while the latter is created with SAS/ACCESS software. Its primary use is to read data from or write data to a DBMS.

SQL views syntax

We will explore SQL views using data for a car dealership. A partial view of the dataset is shown in the following table:

Date	Day	Car	Units	Team	Avg_Price
27JUL2019	Sat	Alpha	25	A1	39450
27JUL2019	Sat	Alpha	23	A2	39850
27JUL2019	Sat	Omega	29	A3	67600
27JUL2019	Sat	Omega	20	A4	68100
28JUL2019	Sun	Alpha	15	A1	39050
28JUL2019	Sun	Alpha	18	A2	39550
28JUL2019	Sun	Omega	19	A3	67900
28JUL2019	Sun	Omega	16	A4	68300

Let's create a sales report that shows the daily sales figures by car type. Apart from changing the `table` keyword to `view`, we don't need to change the syntax of the standard SQL query we have been using:

```
Proc SQL;
  Title 'Sales Report for Management';
  Create View Sales_MI As
Select Date, Day, Car, SUM(Units) As Units_Sold, SUM(Units*Avg_Price) As
Revenue
  From Dealership
  Group by 1,2,3;
Quit;
```

This writes the following message in the LOG:

```
NOTE: SQL view WORK.SALES_MI has been defined.
```

Describing views

Before we move on and explore the output of the view, let's find out about the role of the `Describe` statement. If an analyst has been passed on the name of the view and doesn't have access to the code that was used to create it, they can still see how the view has been constructed. They can see the structure of the view by using the `Describe` statement. The same statement can be used for a table as well:

```
Proc SQL;
   Describe Table Dealership;
   Describe View Sales_MI;
Quit;
```

The following message is written in the LOG:

```
73 Proc SQL;
74 Describe Table Dealership;
NOTE: SQL table WORK.DEALERSHIP was created like:

 create table WORK.DEALERSHIP( bufsize=65536 )
    (
     Date num format=DATE9.,
     Day char(8),
     Car char(8),
     Units num,
     Team char(8),
     Avg_Price num
    );

75 Describe View Sales_MI;

NOTE: SQL view WORK.SALES_MI is defined as:

select Date, Day, Car, SUM(Units) as Units_Sold,
SUM(Units * Avg_Price) as Revenue
         from DEALERSHIP
      group by 1, 2, 3;

76 Quit;
```

Even though the DATA step was used to create the table dealership, the `Describe` statement has the output in the form of a Proc SQL table. The `Describe` statement also writes the information of the index if it's present on the table. If a view is password-enabled, you will have to specify the password before trying to access the definition.

If you have created a view leveraging another view, the `Feedback` option can be used with the `Describe` statement to help us understand the subqueries that were used:

```
Proc SQL;
  Create View Sales_All_Cars As
    Select Date, Day, Sum(Units_Sold) As Units
  From Sales_MI
  Group by 1,2;

  Title 'Use Another View';
  Create View Join As
    Select a.*, b.Units
  From Dealership As a Left Join Sales_All_Cars As b
  On A.Date=B.Date
  And A.Day=B.Day;
Quit;
```

We have created a view that summarizes the sales at a date level by leveraging the first view we connected. We will create another view that joins to the second view we created that has a summary at a date level:

```
Proc SQL Feedback;
  Select * From Join;
Quit;
```

Let's look at how the `Feedback` option helps us decipher the structure of the `Join` view. The following statements are written to the LOG:

```
89 Proc SQL Feedback;
90 Select * From Join;
NOTE: Statement transforms to:
Select A.Date, A.Day, A.Car, A.Units, A.Team, A.Avg_Price, Units
  from ( select A.Date, A.Day, A.Car, A.Units, A.Team,
A.Avg_Price, Units
    from WORK.DEALERSHIP A left outer join
          ( select DEALERSHIP.Date, DEALERSHIP.Day,
SUM(Units_Sold) as Units
      from ( select DEALERSHIP.Date, DEALERSHIP.Day,
DEALERSHIP.Car, SUM(DEALERSHIP.Units) as Units_Sold,
SUM(DEALERSHIP.Units * DEALERSHIP.Avg_Price) as Revenue
          from WORK.DEALERSHIP
      group by 1, 2, 3)
```

```
    group by 1, 2)
  on (A.Date = DEALERSHIP.Date) and (A.Day = DEALERSHIP.Day)
);

91 Quit;
```

The underlying code of the view SALES_MI is also written in the LOG due to the FEEDBACK option used.

Let's go back and explore the results of our original SALES_MI view, which will give us the following Sales MI output:

Date	Day	Car	Units_Sold	Revenue
20JUL2019	Sat	Alpha	40	1562000
20JUL2019	Sat	Omega	47	3171000
21JUL2019	Sun	Alpha	26	1020600
21JUL2019	Sun	Omega	27	1820300
22JUL2019	Mon	Alpha	25	984700
22JUL2019	Mon	Omega	19	1287600
23JUL2019	Tue	Alpha	28	1100300
23JUL2019	Tue	Omega	19	1289400
24JUL2019	Wed	Alpha	17	669500
24JUL2019	Wed	Omega	27	1818300
25JUL2019	Thu	Alpha	34	1337800
25JUL2019	Thu	Omega	30	2040500
26JUL2019	Fri	Alpha	37	1469000
26JUL2019	Fri	Omega	42	2856400
27JUL2019	Sat	Alpha	48	1902800
27JUL2019	Sat	Omega	49	3322400
28JUL2019	Sun	Alpha	33	1297650
28JUL2019	Sun	Omega	35	2382900

Using this view, we would have been able to create a daily sales report stating the units sold and the revenue that's been generated for each car.

What would happen if, instead of creating a SALES_MI view, we wanted to create a DEALERSHIP view?

You will get a message similar to the following error message if you try to create a view that has the name of an existing dataset:

```
ERROR: Unable to create WORK.DEALERSHIP.VIEW because WORK.DEALERSHIP.DATA
already exists.
```

However, if you try to create a view that has been previously named and defined, SAS will not throw an error, and you will be able to overwrite the previously defined view.

Optimizing performance using views

To showcase the optimization of queries using views, let's create a big dataset. The steps to optimize performance using views are as follows:

1. Use the following code to create the biggest dataset we have used in this book so far:

```
Data Dealership_Looped;
   Do i = 1 to 1000000;
      Do j = 1 to n;
         Set Dealership Nobs=n Point=j;
         Output;
         End;
      End;
   Stop;
Run;
```

We have leveraged the dealership data and looped it a million times to create a dataset with 36 million records.

2. Use the following code to create a table where 1 new column is added and then Proc Means is run on it:

```
Proc SQL;
  Create Table Multiplier As
    Select *, Avg_Price*1.5 As Avg_new
  From Dealership_Looped;
Quit;
Proc Means Data = Multiplier;
Var Avg_New;
Run;
```

The time that's needed to create the table and run proc means on it is written to the LOG:

```
NOTE: Table WORK.MULTIPLIER created, with 36000000 rows and 8
columns.

NOTE: PROCEDURE SQL used (Total process time):
      real time 36.48 seconds
      cpu time 33.13 seconds

NOTE: There were 36000000 observations read from the data set
WORK.MULTIPLIER.

NOTE: PROCEDURE MEANS used (Total process time):
      real time 25.34 seconds
      cpu time 25.01 seconds
```

3. As an alternative, you will have to create a view with the new column and run proc means on it:

```
Proc SQL;
   Create View Multiplier As
      Select *, Avg_Price*1.5 As Avg_new
   From Dealership_Looped;
Quit;


Proc Means Data = Multiplier_Alt;
Var Avg_New;
Run;
```

Let's look at the time that's needed to create the view and run proc means on it:

```
NOTE: SQL view WORK.MULTIPLIER_ALT has been defined.

NOTE: PROCEDURE SQL used (Total process time):
      real time 0.01 seconds
      cpu time 0.01 seconds

NOTE: There were 36000000 observations read from the data set
WORK.DEALERSHIP_LOOPED.
NOTE: There were 36000000 observations read from the data set
WORK.MULTIPLIER_ALT.
NOTE: PROCEDURE MEANS used (Total process time):
      real time 38.21 seconds
      cpu time 40.38 seconds
```

Using the data table, proc means took a total of 62.22 seconds to complete, while it only took 38.22 seconds using views. That's a saving of 61% when using views. Most SAS users ignore views and, as a result, miss out on the huge savings that can be made on large datasets. This was a simple example where the bigger dataset had a few columns and wasn't joined to multiple datasets, and a simple variable was created before running proc means. A complex operation on a table would have led to a lot of CPU usage, and so using views would have made even more sense.

The reason we saved time is that a substantial amount of time was spent on the table's creation. The `dealership_looped` dataset, which had 36 million rows, was read into memory. When proc means was run again, 36 million rows were read from the multiplier table. In the case of view creation, we read the 36 million rows from `dealership_looped`. However, when proc means was run, the underlying view referred to the same 36 million rows that were read while creating the view and did not force proc means to do a disk I/O. This meant that running proc means took much less time using the view.

Making changes with Proc SQL

Proc SQL offers users the option of deleting, altering, and updating records. Along with looking at examples of how to do this in Proc SQL, we will also talk about alternatives that are available in SAS. We will use our 36 million record dataset to also try to understand the time that was taken to complete these tasks. Although the number of records is high, please beware that it is a long table, not a wide one. In practice, tables will be wider, and the time that it would take to process a typical table with such a structure may be longer.

Deleting

The general form of the DELETE statement is as follows:

```
DELETE FROM table - name
  <WHERE expression>;
```

Here, we have the following:

- The table name is the table whose records need to be deleted
- WHERE is an optional statement if only certain records from the table need to be deleted

We will try to subset the data before deleting it. Let's delete all the records that were created due to the looping exercise on the dealership data.

Run the following code:

```
Proc Sql;
  Delete From Dealership_looped;
                  Where i gt 1;
  Quit;
```

Did you notice anything wrong in the preceding syntax? The semicolon is in the wrong place. Due to the semicolon before the WHERE clause, all the records will be deleted from the table. SAS won't throw an error, but after deleting the records, it will put a NOTE in the LOG stating that the WHERE clause isn't supported in this manner in SAS. Therefore, be careful and don't put a semicolon after the dataset name unless you don't have a WHERE condition.

Let's remove the semicolon after the dataset name. We will get the following LOG message:

```
NOTE: 35999964 rows were deleted from WORK.DEALERSHIP_LOOPED.

NOTE: PROCEDURE SQL used (Total process time):
      real time 2:32.45
      cpu time 2:29.17
```

The same DELETE statement can also help delete the records from the underlying table specified in the view.

There is an alternate way to delete the records using the built-in procedure. Let's compare the time it takes to complete the task using PROC DELETE. The Proc SQL delete syntax was run without the WHERE condition so that the time that was taken was specified:

```
NOTE: 36000000 rows were deleted from WORK.DEALERSHIP_LOOPED.

 75 Quit;
NOTE: PROCEDURE SQL used (Total process time):
      real time 1:31.46
      cpu time 1:29.89
```

Deleting the dataset using PROC DELETE only took 0.16 seconds, which is considerably lower than the 1:31:46 that we had by using Proc SQL without the WHERE condition:

```
NOTE: Deleting WORK.DEALERSHIP_LOOPED (memtype=DATA).
 NOTE: PROCEDURE DELETE used (Total process time):
      real time 0.16 seconds
      cpu time 0.14 seconds
```

The code that we used for PROC DELETE was as follows:

```
Proc Delete Data=Dealership_looped;
Run;
```

We can delete multiple datasets using the preceding procedure. The dataset names can be specified as follows:

```
Data = Lib1.A Lib2.B Lib.C (genum = all);
```

Specifying the `genum` option deletes all historical versions of the datasets.

We can also delete the datasets using the DATASETS procedure:

```
Proc Datasets Library=WORK;
  Delete Dealership_Looped;
Run;
```

The preceding code also took approximately the same time as using the DELETE procedure.

Note that the PROC DELETE option doesn't allow for a conditional delete. A warning is produced in the following LOG message to that effect. However, what isn't stated in the LOG is that the `dealership_looped` dataset has already been deleted:

```
73 Proc Delete Data=dealership_looped;
 74 Where i gt 1;
 WARNING: No data sets qualify for WHERE processing.
```

If you are trying to delete a dataset, I recommend going for the built-in DELETE or DATASETS procedures. However, if a subset of the data needs to be deleted, try using Proc SQL with the WHERE option.

Altering

Using the ALTER statement, we can drop (delete) columns, modify columns, and add columns. The general form of this statement is as follows:

```
ALTER TABLE table-name
       <DROP column-name-1<, ... column-name-n>>
       <MODIFY column-definition-1<, ... column-definition-n>>;
       <ADD column-definition-1<, ... column-definition-n>>
```

Here, `table-name` specifies the name of the table in which columns will be added, dropped, or modified.

At least one of the following clauses must be specified:

- DROP specifies one or more column-names for columns to be dropped.
- MODIFY specifies one or more column-definitions for columns to be modified, where column-definition specifies a column to be added or modified, and is formatted as follows: `column-name data-type <(column-width)>` `<column-modifier-1 < ...column-modifier-n>>`.
- ADD specifies one or more column-definitions for columns to be added.

We will drop the i column from the `Dealership` table. This is a looping variable that offers no meaning to the data. We only retained it to showcase the DELETE option using Proc SQL:

```
Proc Sql;
  Alter Table Dealership_Looped
    Drop i;
Quit;
```

The LOG message doesn't clearly state the action that has been performed:

```
NOTE: Table WORK.DEALERSHIP_LOOPED has been modified, with 6 columns.
```

In this case, the analyst needs to remember that the `Dealership` table had six columns and that one column (i) was added while looping. Hence, removing one column still leaves us with the original count of six columns. The partial view of the table before dropping i is as follows:

i	Date	Day	Car	Units	Team	Avg_Price
1	20JUL2019	Sat	Alpha	20	A1	39000
1	20JUL2019	Sat	Alpha	20	A2	39100
1	20JUL2019	Sat	Omega	25	A3	67000
1	20JUL2019	Sat	Omega	22	A4	68000
1	21JUL2019	Sun	Alpha	12	A1	39200
1	21JUL2019	Sun	Alpha	14	A2	39300
1	21JUL2019	Sun	Omega	16	A3	67500
1	21JUL2019	Sun	Omega	11	A4	67300
1	22JUL2019	Mon	Alpha	14	A1	39300
1	22JUL2019	Mon	Alpha	11	A2	39500

The partial view of the table after dropping i is as follows:

Date	Day	Car	Units	Team	Avg_Price
20JUL2019	Sat	Alpha	20	A1	39000
20JUL2019	Sat	Alpha	20	A2	39100
20JUL2019	Sat	Omega	25	A3	67000
20JUL2019	Sat	Omega	22	A4	68000
21JUL2019	Sun	Alpha	12	A1	39200
21JUL2019	Sun	Alpha	14	A2	39300
21JUL2019	Sun	Omega	16	A3	67500
21JUL2019	Sun	Omega	11	A4	67300
22JUL2019	Mon	Alpha	14	A1	39300
22JUL2019	Mon	Alpha	11	A2	39500

The dataset now includes variables that are of importance when we wish to prepare the sales MI.

If we want to modify a column using the ALTER statement, we can perform the following changes to a column:

- Informat
- Format
- Label
- Length of a character variable

In our `Dealership` table, the average price is in USD. We will specify a format by modifying the existing variable. We will also change the width of the `Car` variable as we anticipate that another car with a longer name will be added to the MI report.

Currently, the properties of the `Dealership` table are as follows:

General	Columns	Extended Attributes	Column Extended Attributes		
Column Name	**Type**	**Length**	**Format**	**Informat**	**Label**
DATE	Numeric	8	DATE9.		
DAY	Char	8			
CAR	Char	8			
UNITS	Numeric	8			
TEAM	Char	8			
AVG_PRICE	Numeric	8			

We will use the following code to modify the columns:

```
Proc Sql;
   Alter Table Dealership
Modify Car char(12),
Avg_Price format=Dollar11.2 label="Avg Price USD";
Quit;
```

We have specified the new length of the `Car` variable and changed the format of the `Avg_Price` variable. Furthermore, we have added a label to the variable. The results are visible in the new properties for the table:

General	Columns	Extended Attributes	Column Extended Attributes		
Column Name	**Type**	**Length**	**Format**	**Informat**	**Label**
DATE	Numeric	8	DATE9.		
DAY	Char	8			
CAR	Char	12			
UNITS	Numeric	8			
TEAM	Char	8			
AVG_PRICE	Numeric	8	DOLLAR11.2		Avg Price USD

Remember to use a comma to separate the variables in the ALTER statement. A semicolon shouldn't be used after the ALTER statement until you have finished specifying the changes you wish to make.

Another aspect of the ALTER statement that we discussed is adding columns to a table. The MI report will be used by management to determine the performance rating for each team. Then, they can decide on the incentive pool for the team. We will add a couple of columns to the table using the following code:

```
Proc Sql;
  Alter Table Dealership
    Add Rating char(3),
      Incentive num format=Dollar11.2;
Quit;
```

No formula or logic has been added to the columns. The following table is a partial view of the Altered Dealership Table With New Columns:

Date	Day	Car	Units	Team	Avg_Price	Rating	Incentive
20JUL2019	Sat	Alpha	20	A1	39000		.
20JUL2019	Sat	Alpha	20	A2	39100		.
20JUL2019	Sat	Omega	25	A3	67000		.
20JUL2019	Sat	Omega	22	A4	68000		.
21JUL2019	Sun	Alpha	12	A1	39200		.
21JUL2019	Sun	Alpha	14	A2	39300		.
21JUL2019	Sun	Omega	16	A3	67500		.
21JUL2019	Sun	Omega	11	A4	67300		.
22JUL2019	Mon	Alpha	14	A1	39300		.
22JUL2019	Mon	Alpha	11	A2	39500		.

Apart from dropping, modifying, and adding columns, the ALTER statement can also be used to add/drop integrity constraints to/from an existing table. The integrity constraints and their default names are as follows:

Default Name	Constraint Type
NMxxxx	Not null
UNxxxx	Unique
CKxxxx	Check
PKxxxx	Primary key
FKxxxx	Foreign key

The clause to add or drop an integrity constraint in the ALTER statement is as follows:

```
<ADD constraint-specification-1 <, constraint-specification-2, ...>>
<DROP CONSTRAINT constraint-name-1 <, constraint-name-2, ...>>
<DROP FOREIGN KEY constraint-name>
<DROP PRIMARY KEY>
```

There are some aspects to bear in mind when using the ALTER statement:

- Renaming is not possible. To rename a column, use the DATA step or the Proc SQL step.
- If a column has an index defined, altering the value of the column will still keep the index that is defined for it. If a column is dropped, then all the indexes where the column is are dropped.
- When a column is added to the table using the ALTER statement, missing values are assigned to the column. To assign values, use the UPDATE statement in conjunction with the ALTER statement.

Identifying duplicates using Proc SQL

The simplest way to remove duplicates in Proc SQL is by using the Distinct statement. We will use it on the Dealership_Looped dataset, where the i column, which is used as a looping counter, has been dropped:

```
Proc Sql;
  Create Table Distinct_Dealership_Looped As
      Select Distinct *
    From Dealership_Looped
  ;
Quit;
```

Using the `Distinct` statement, we have correctly identified the duplicates we created as part of the DO LOOPS. We are now left with the original number of 36 records we had. This can be confirmed by looking at the following LOG:

```
NOTE: Table WORK.DISTINCT_DEALERSHIP_LOOPED created, with 36 rows and 6
columns.

NOTE: PROCEDURE SQL used (Total process time):
      real time 1:56.01
      cpu time 1:05.78
```

Let's find out how would we have fared in terms of runtime if we had used PROC SORT. After all, PROC SORT is the most popular way among analysts to remove duplicates. At times, some analysts struggle with computing resources and execute PROC SORT on large databases. However, instead of relying on alternative duplicate removal aspects, analysts tend to use WHERE conditions, indexes, or some other means to try to execute their queries. However, let's discuss this aspect in a bit. First, we will run PROC SORT to identify our duplicates:

```
Proc Sort Data = Dealership_Looped Nodupkey;
By _All_;
Run;
```

There's not much difference between the Proc SQL and PROC SORT identification of duplicates in terms of runtime. It is worth noting that the underlying data was presorted and that we didn't utilize any resources for that action since sorting is a mandatory condition of running PROC SORT:

```
NOTE: There were 36000000 observations read from the data set
WORK.DEALERSHIP_LOOPED.
NOTE: 35999964 observations with duplicate key values were deleted.
NOTE: The data set WORK.DEALERSHIP_LOOPED has 36 observations and 6
variables.
NOTE: PROCEDURE SORT used (Total process time):
      real time 1:47.06
      cpu time 1:05.42
```

Let's take another look at the `Class` dataset we used in the previous chapters. Have a look at observations 1, 3, 5, 6, 11, and 12, which have `Height` values that are similar in two rows of data:

Obs	ClassID	Year	Age	Height	Weight
1	A1234	2013	8	85	34
2	A2323	2013	9	81	36
3	B3423	2013	8	80	31
4	B5324	2013	9	70	35
5	C2342	2013	9	80	31
6	D3242	2013	9	85	30
7	A1234	2019	14	105	64
8	A2323	2019	15	101	66
9	B3423	2019	14	100	61
10	B5324	2019	15	90	55
11	C2342	2019	15	112	70
12	D3242	2019	14	112	70

If we want to treat these similar values as duplicates at a key level for the `Height` and `Age` variables, we can identify and remove these records using Proc SQL:

```
Proc Sql;
  Create Table Class_NoDuP_Height As
    Select Distinct(Height), ClassID, Year, Age, Weight
      From Class
        Group by Height
        Having Age = Max(Age);
Quit;
```

The following output is produced:

Obs	Height	ClassID	Year	Age	Weight
1	70	B5324	2013	9	35
2	80	C2342	2013	9	31
3	81	A2323	2013	9	36
4	85	D3242	2013	9	30
5	90	B5324	2019	15	55
6	100	B3423	2019	14	61
7	101	A2323	2019	15	66
8	105	A1234	2019	14	64
9	112	C2342	2019	15	70

Where the records had similar height, the one with the higher age has been retained. You can add various combinations to serve as keys and identify and remove duplicates based on these keys. The variables that are part of the key need to be mentioned in the HAVING statement.

We can obtain similar results using the SUMMARY procedure, as follows:

```
Proc Summary Data=Class Nway;
Class Height;
Id ClassID Year Age Weight;
Output Out=Class_without_DupKey (Drop=_type_);
Run;
```

The following output is produced:

Height	ClassID	Year	Age	Weight	_FREQ_
70	B5324	2013	9	35	1
80	C2342	2013	9	31	2
81	A2323	2013	9	36	1
85	D3242	2013	9	30	2
90	B5324	2019	15	55	1
100	B3423	2019	14	61	1
101	A2323	2019	15	66	1
105	A1234	2019	14	64	1
112	D3242	2019	14	70	2

In this section, we have looked at three different ways of removing duplicates. The SORT procedure is a heavy resource method for removing duplicates. Do try to consider the alternative methods when it comes to removing records, even though they might not be the most intuitive methods for you if you already have been using SAS for a while. Let's go back to the original problem of deduping the `Dealership_looped` dataset and see how the SUMMARY procedure performs at runtime:

```
Proc Summary Data = Dealership_Looped Nway;
  Class _All_;
    Output Out = Proc_Summary_Nway;
Run;
```

What's notable here is that Proc SQL took 1:56:01 to complete and that PROC SORT took 1:47:06 to remove the duplicates, but the SUMMARY procedure took substantially less time, that is, 00:21:97, to remove the duplicates from the `Dealership_looped` dataset:

```
NOTE: PROCEDURE SUMMARY used (Total process time):
      real time 21.97 seconds
      cpu time 33.25 seconds
```

Once again, this substantiates the view that PROC SORT may not always be the best tool for an analyst when it comes to removing duplicates.

Creating an index in Proc SQL

As we have discovered earlier, creating an index is an efficient way of dealing with large datasets. Proc SQL also offers us the option to create and manage indexes.

An index is an auxiliary file that is defined on one or more variables, which are called key columns. The index may be primary or composite, that is, formed of one or multiple variables. The index stores the unique column values and directions that allow access to rows in an indexed manner. Proc SQL benefits from indexes by reading the required record directly rather than following the sequential method. Hopefully, this recap on indexes will bring back memories of the examples we used as part of the DATA step.

Let's create a simple and composite index. The datasets in this chapter are smaller and don't contain hundreds of unique values. There will be no significant benefit of creating an index on such datasets. However, the following blocks of code will help you define indexes across any datasets you encounter:

```
Proc SQL;
  Create Index Team
    On Dealership(Team);
```

```
Create Index Composite
   On Dealership(Car, Team);
Quit;
```

We get confirmation that the index was created in the following LOG:

```
NOTE: Simple index Team has been defined.
NOTE: Composite index Composite has been defined.
```

Apart from checking the LOG, we can use the DESCRIBE statement, which gives us details about the indexes on a table. We will use it to describe the Dealership table:

```
Proc SQL;
  Describe Table Dealership;
Quit;
```

We get the following message in the LOG:

```
create table WORK.DEALERSHIP( bufsize=65536 )
   (
    Date num format=DATE9.,
    Day char(8),
    Car char(8),
    Units num,
    Team char(8),
    Avg_Price num
   );
  create index Composite on WORK.DEALERSHIP(Car,Team);
  create index Team on WORK.DEALERSHIP(Team);
```

Another way we can check if an index is being used is by specifying the MSGLEVEL option:

```
Options Msglevel=i;
Proc SQL NoPrint;
  Select *
    From Dealership
      Where Team = 'A4';
Quit;
```

 Note that this option doesn't list all the indexes that are defined on the table.

This produces the following message in the LOG:

```
INFO: Index Team selected for WHERE clause optimization.
```

If you know the index name, you can also specify it in your query:

```
Proc SQL NoPrint;
  Select *
    From Dealership (idxname = Team)
      Where Team = 'A4';
Quit;
```

There may be an instance where you want SAS to ignore the indexes. You can do so with the following code:

```
Proc SQL NoPrint;
  Select *
    From Dealership (idxwhere = no)
      Where Team = 'A4';
Quit;
```

This will produce the following LOG message:

```
INFO: Data set option (IDXWHERE=NO) forced a sequential pass of the data
rather than use of an index for where-clause processing.
```

Finally, to drop the indexes from the `Dealership` table, we will use the following code:

```
Proc SQL NoPrint;
  Drop Index Team, Composite
    From Dealership;
Quit;
```

Macros and Proc SQL

In the previous chapter, we spoke about learning about the essentials of Proc SQL before turning our attention to Macros. Let's explore this aspect in Proc SQL since some of the functionalities of macros take longer to code using the DATA step.

Macros in Proc SQL are powerful, primarily due to the INTO clause that is offered in conjunction with the SELECT statement. The INTO clause cannot be used while creating tables or views. The INTO clause for the SELECT statement can assign the result of a calculation or the value of a data column (variable) to a macro variable. If the macro variable doesn't exist, INTO creates it. You can use PROC SQL's SQLOBS macro variable to see how many rows (observations) were produced by a SELECT statement. Let's explore this clause using a few examples.

Creating a macro variable using Into

We know that our `Dealership_Looped` table has 36 million observations. What if we wanted to store this value in a macro variable? Based on our understanding of macros from previous chapters, we will write the following command to create a macro variable with the value:

```
%Let n = 36000000;
```

In Proc SQL, the alternative is as follows:

```
Proc SQL;
   Select Count(*) Into: n_sql
      From Dealership_Looped;
Quit;
```

If we run the preceding commands, then the value is 36 million:

```
%Put n_sql is &n_sql;
n_sql is 36000000
```

The INTO clause is quite helpful because, without prior information on the row count, we could still create a macro variable that holds the value of the row count. This clause comes in handy when we are writing a macro definition where the dataset population is dynamic.

Creating multiple macro variables using Into

We aren't restricted to creating a single macro variable using INTO. Let's store the names of the various teams in our `Dealership` table into macro variables:

```
Proc SQL;
   Select Distinct Team Into: Teams
      From Dealership;
Quit;
```

Running the preceding command produces the following output in the display window:

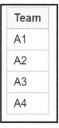

Team
A1
A2
A3
A4

The output seems to suggest that the macro variable that was formed has four values, and so we might have multiple macro variables holding the name of each team. However, this isn't the case. The output that's generated has the variable name `Team`, whereas we requested the `Teams` macro variable. The preceding code has only produced a single macro variable, that is, the first value of the team it found.

Incidentally, while creating a macro definition, you might want to suppress the output being produced every time an INTO clause is used. Just use the `NOPRINT` option. Meanwhile, we can establish that only one macro variable was created by running the following statement and looking at the value in the LOG. Remember, the LOG tells you a lot about what might have actually happened while executing the code:

```
%Put Teams is &Teams;
Teams is A1
```

So, what went wrong? We never specified the names of the macro variables we wanted. Let's correct our mistake:

```
Proc SQL NoPrint;
  Select Distinct Team Into: Team1- :Team4
    From Dealership;
Quit;

%Put Team1 is &Team1;
%Put Team2 is &Team2;
%Put Team3 is &Team3;
%Put Team4 is &Team4;
```

This produces the following LOG message:

```
Team1 is A1
Team2 is A2
Team3 is A3
Team4 is A4
```

What if we didn't want the output in a results window and, instead of the macro variable only having a value of `A1`, we wanted just one macro variable with all the possible values of the `Team` variable? In that case, run the following query:

```
Proc SQL NoPrint;
  Select Distinct Team Into: Teams Separated by ","
    From Dealership;
Quit;

%Put Teams is &Teams;
```

The following message appears in the LOG:

```
Teams is A1,A2,A3,A4
```

Let's combine what we have learned about INTO and create a macro definition that will produce the looped `Dealership_looped` dataset for us. The number of loops will be driven by the count of the rows in the base `Dealership` dataset:

```
%Macro Loop_Dealership_Table;

Proc SQL NoPrint;
   Select Count(*) Into: Count
From Dealership;
Quit;

Data Dealership_Looped (Drop=i);
   do i = 1 to 1000000;
      do j = 1 to &Count;
         set Dealership nobs=n point=j;
         output;
         end;
      end;
   stop;
Run;

%Mend;

%Loop_Dealership_Table;
```

The preceding macro will produce the `Dealership_Looped` dataset, which has 36 million records.

Summary

We started off this chapter by exploring the concept of views. We discussed multiple advantages of views, including their core strengths of protecting sensitive information, easing the coding process, and giving us the ability to use the most recent version of the source data. Apart from these core advantages, we also explored how views can help optimize queries. Manipulating data was also a central theme of this chapter as we looked at deleting columns and rows, modifying columns, and adding columns. We also looked at the identification of duplicates as part of manipulating data. Here, we compared Proc SQL to the common deduplication technique of Proc SORT and also explored the not-so-common use of the SUMMARY procedure for this task.

While we covered indexes as a topic previously in this book, we looked at creating, controlling, monitoring, and deleting indexes. Toward the end of this chapter, we looked at macros in Proc SQL, where we discovered the all-important INTO statement. We created a macro definition that highlighted the versatility of the INTO statement. This chapter should help you feel confident in writing advanced Proc SQL code.

In the next chapter, we will learn about data visualization in SAS.

Section 5: Data Visualization and Reporting

5

This part provides a discussion of various reporting functions and output styles. It helps the reader to present a data story that provides end-mile connectivity and helps to showcase the insights as a result of analyzing data.

This section comprises the following chapters:

- Chapter 9, *Data Visualization*
- Chapter 10, *Reporting and Output Delivery System*

Data Visualization 9

We have run various procedures and utilized countless functions in the previous chapters. All of them produced some kind of output that was represented as tables or charts. Visualization of the output would have played an important role in developing your understanding of what SAS has to offer. As a data user, you will want to tell your data story using visuals. Some of the built-in SAS procedures produce graphs by default. However, in this chapter, we will look at producing charts and controlling various facets of their output options. There is no single or right way of visualizing the data. By the end of this chapter, you will have gained insight into data visualization using a variety of techniques. This should allow you to tell your data story in your preferred way. While there are many more charts that can be built into SAS, we will focus on a few key ones that should help deliver the message your data wants to.

In this chapter, we will cover the following topics:

- The role of data visualization in analytics
- Histograms
- Line plots
- Vertical and horizontal bar charts
- Scatter charts
- Box plot

The role of data visualization in analytics

A picture is worth a thousand words. You must have heard this adage before. This adage is relevant in the world of data analytics. The importance of data visualization isn't ebbing but is on the rise considering that most institutions have big data but are grappling with how to best use it. One of the major hurdles in using big data to its full potential is figuring out what the underlying data is telling you. To convey insights from the data, visualization plays a key role. Only the summary of the data is visualized. If this visual sends a strong message about the data, chances are some action will be taken based on the visual. Let's list a few reasons for the visualization of data:

- Information can be absorbed quickly using data.
- It helps us understand trends and patterns.
- Audiences can usually relate to graphics more easily than numbers in tabular formats.
- It can hold the audience's interest for long.
- It helps us identify outliers easily with the aid of a few specific charts (for example, box plots, and scatter plots).
- It moves the discourse away from the methodology of collecting and analyzing data to inferring the message.

Histograms

The exact use of histograms is to assess the probability distribution of a given variable by plotting the frequencies of observations occurring in certain ranges of values. They were first described by Karl Pearson. In their most simplistic form, histograms plot the frequency of a variable in a range of values called bins. We have chosen to start this chapter by describing histograms as they are the simplest of graphs that only accommodate one variable. Adding a density curve makes them a bit more informative but let's start with the basic form of the histogram. We will use the `Class` dataset that has been extensively used in the previous chapters:

```
Proc SGPLOT Data = Class;
   Histogram Height;
   Title 'Height of children in class across years';
Run;
```

The only change to the `Class` dataset is that the `Weight` variable has been renamed `Weights` in this chapter.

This produces the following diagram:

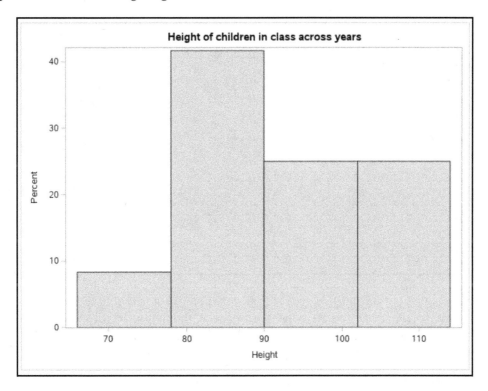

Out of the 12 observations in the dataset, one child has a height equal to 70. The observation for this child is included in the first bin. The observation contributes to around 8% of the total number of observations. As per the preceding diagram, around 40% of observations have a height between 80 to 90 units. Looking at the spread of the height variable in the dataset, the height of 70 for the ClassID B5324 looks to be an outlier. Let's remove the observation using the following code:

```
Proc SGPLOT Data = Class;
  Histogram Height;
  Title 'Height of children in class across years';
  Where ClassID ne 'B5324';
Run;
```

We get the following diagram as the output:

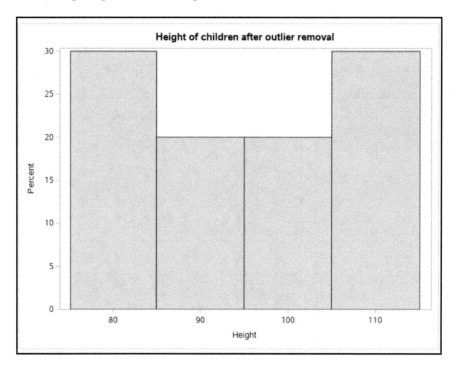

While we removed the outlier in the overall population, we know that the `Class` dataset has observations from 2013 and 2019. It's definite that, the height of students between those six years would have changed, given the fact that these are school-going children. We will use the following code to create a panel of histograms for both observation years:

```
Proc SGPANEL Data=Class;
   Panelby Year / Rows=2 Layout=Rowlattice;
   Histogram Height;
Run;
```

The following diagram clearly illustrates the difference in heights across the observation years:

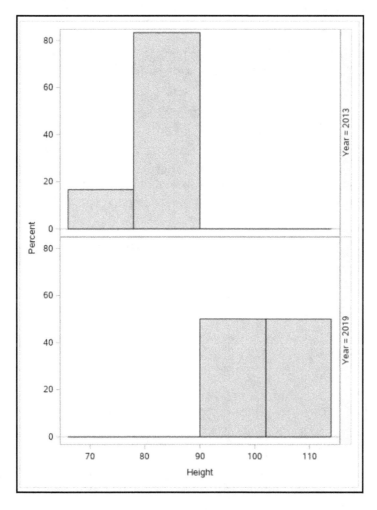

The `Panelby` statement helps create the histograms in the panel. `Rows` specifies the number of groups we have in the data. In our dataset, the heights do not overlap across the years. The growth spurt seems to have come to an end and the children don't seem to be getting taller when you look at the lower panel. However, that's not true. The *x*-axis of the year 2019 has height values over a large band clubbed together in two bins. Both bins have the same frequency of 50%. However, this doesn't mean that the height of children in the year 2019 is the same.

The visualization of data should aid in the correct interpretation of the data. You won't be able to present both the data and the graph to your end users on certain occasions. The visualization of the data should make drawing inferences and insights a relatively easier task.

One way to make the histogram easy to interpret is to control the number and size of the bins on the *x*-axis. You can use the following code to achieve that:

```
Proc SGPLOT Data = Class;
   Histogram Height / Binstart=70 Binwidth=.5 Scale=count;
   Title 'Height of Class in Customized Bins';
Run;
```

The following diagram is an extreme example where we have created a bin that is so small in range that we have ended up with the number of bins equal to the number of observations. We could have also used the NBins option to control the number of bins:

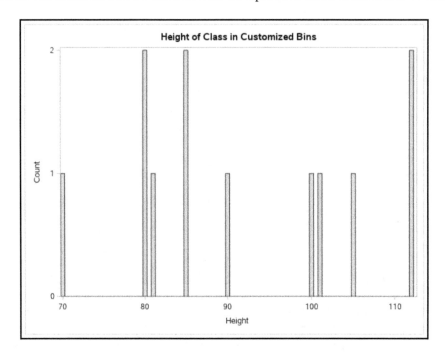

While defining histograms, we mentioned that they are useful if we wish to understand the probability distribution of the variable.

Let's plot a density curve, which, in the case of SGPLOT, is the normal density curve by default. The parameters are estimated from the data:

```
Proc SGPLOT Data = Class;
    Histogram Height;
    Density Height;
    Title 'Height of children in class across years';
Run;
```

We have managed to plot the following histogram and density curve together in the same graph:

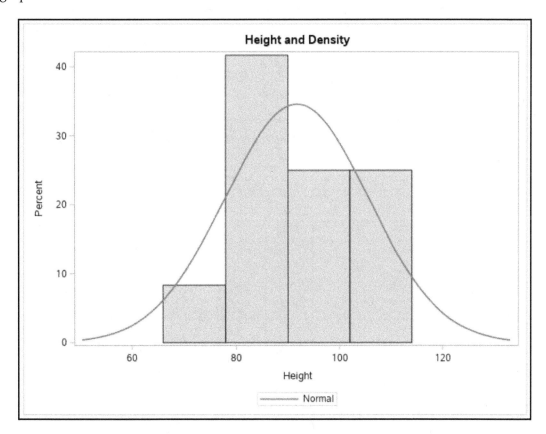

While this chapter isn't about probability distribution functions, we will look at an example where the histogram allows us to plot multiple density functions:

```
Proc SGPLOT Data = Class;
    Histogram Height;
    Density Height;
```

```
   Density Height / Type= Kernel;
   Keylegend / Location = Inside Position = TopRight
   Across = 1 Title = 'Density Curves';
   Title 'Height and Density with Multiple Curves';
Run;
```

The type option can help us specify the density function. The options that are available are normal or kernel. The normal option specifies a normal distribution based on the mean and the standard deviation. The kernel option specifies a non-parametric kernel density estimate.

The Keylegend statement provides a legend to understand which curves have been plotted. There are various options such as location, position, across, and title available in the statement. The location can be inside or outside of the panel of the chart. If you specify the down option instead of across, you will get a horizontal list of the density functions instead of the stacked-up list of density function names that can be seen in the following diagram:

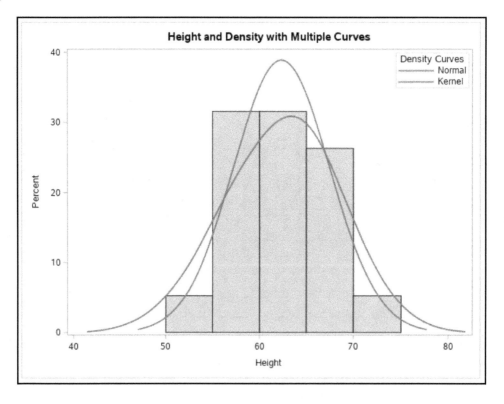

Previously, we created two histograms in the same chart but different panels. But how can you plot their density functions while still being able to compare the histograms? Use the following code to create histograms by group in the same panel and create multiple density plots:

```
Proc SGPLOT Data=Class;
    Histogram Height / Group=Year Transparency=0.5;
    Density Height / Group=Year;
Run;
```

This produces the following output:

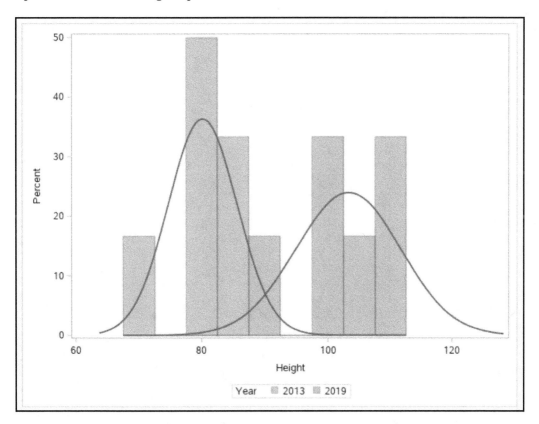

We can create histograms by group in the same panel and create multiple density plots as well.

Line plots

We will explore the most basic format of the line plot on a single axis and move on to exploring a few more aspects of this chart in this section. Use the following query to create a line chart where we are interested in finding out the frequency of the Age variable:

```
Proc SGPLOT Data=Class;
   Vline Age;
   Title 'Basic Form of Line Chart using SGPLOT';
Run;
```

We get the following plot as the output:

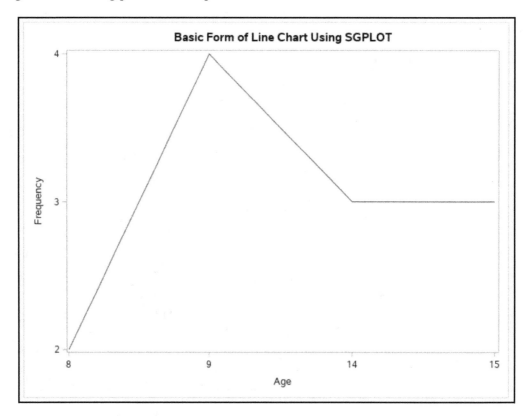

There are only four data points for `Age`, and the *y* axis has the frequency of each data point.

The preceding chart only contains one series. We will not get a meaningful output if we use the following similar code for the `Class` dataset. Let's use the `Cost_Living` dataset we first used in `Chapter 1`, *Introduction to SAS Programming*, to plot a line chart with multiple series:

```
Proc SGPLOT Data=Cost_living;
   Series X=City Y=Index / Legendlabel="Current Yr Index";
   Series X=City Y=Prev_Yr_Index / LEGENDLABEL="Previous Yr Index";
   YAxis Label="Current vs Previous Index";
   Title 'Multiple Series in Line Chart';
Run;
```

In the following output, we get two series plotted on the *y*-axis for each **City** that represents the *x*-axis:

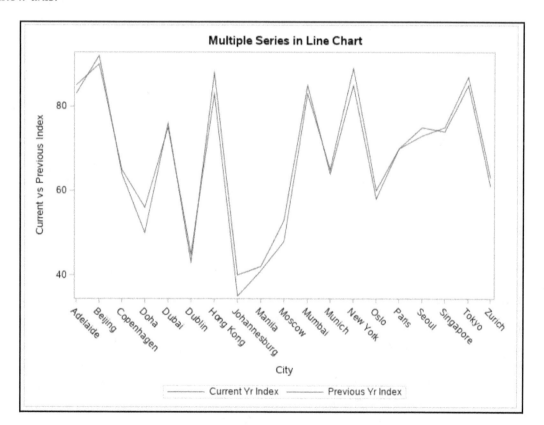

We still haven't utilized the second *y*-axis or the *z*-axis to plot any series. Grouping two series by a third series can allow us to do that, as shown in the following code:

```
Proc SGPLOT Data=Class;
   Vline Age / Response=Height Stat=Mean Markers;
   Vline Age / Response=Weights Stat=Mean Markers Y2AXIS;
   Title 'Age with Response Variables';
Run;
```

This will give us the following plot as output:

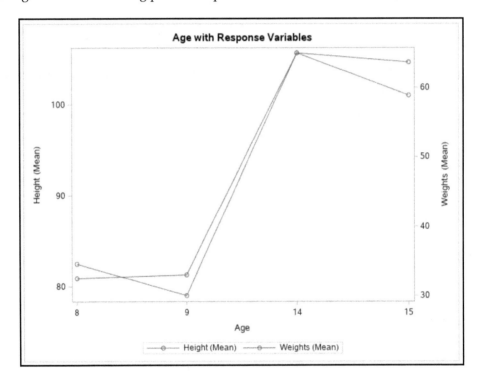

From the output, we can infer that, even though **Age** is increasing from 8 to 9 and then from 14 to 15, the mean height is going down. This is probably due to an observation(s) that has less height compared to other observations. This outlier observation is pulling the mean down. When we look at the data to spot the outlier, we can see that ClassID B5324 has a height of 70 units at age 9. In the same age period, other ClassIDs have a height of 80, 81, and 85 units. The same ClassID, 5324 has a height of 90 units at age 9 whereas, in the same age period, other ClassIDs have a height of 101 and 112 units. The ClassID in question is pulling down the average value and this can be clearly seen in the preceding diagram.

Up until now, we have looked at various charts but none of them answer the question of the increase in height or weight of a student in a simplistic manner. To create this missing piece of visualization, we will sort our `Class` dataset to ensure that we have ordered it by `ClassID`, `Year`, and `Height`:

```
Proc Sort Data = Class Out=Delta;
   By ClassID Year Height;
Run;
```

Now, we will simply run `Proc SGPLOT` by specifying the *x*-axis and *y*-axis and grouping the variables by `ClassID`:

```
Proc SGPLOT Data=Delta;
   Series X=Year Y=Height / Group=ClassID;
   Title 'Change in Height';
Run;
```

This will give us the following plot as the resultant output:

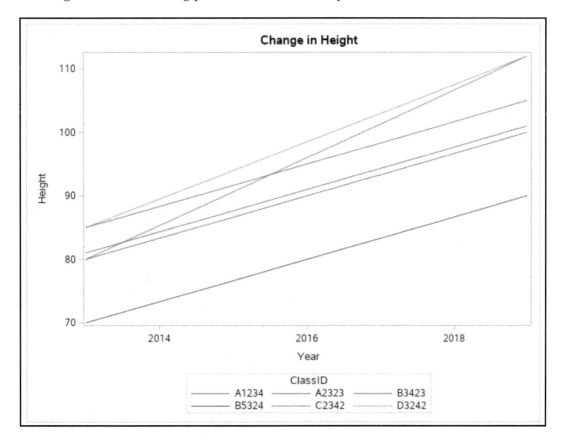

What we have achieved in the preceding chart is a simple visual representation of the change in the height of each student in the class.

Vertical and horizontal bar charts

A vertical bar chart is not a histogram. Remember that the first chart you saw in this chapter was a histogram and its *y*-axis totaled 100%. This won't necessarily happen in every vertical bar chart. A histogram is more than a vertical representation of data, as we saw when we used one to understand the probability distribution function using density curves. Let's delve into how vertical bar charts can make our data visually appealing. As always, we will start with a simple example:

```
Proc SGPLOT Data=Class;
  VBar Height;
  Title ' Basic Form of Vertical Chart';
Run;
```

The chart that's produced is as follows:

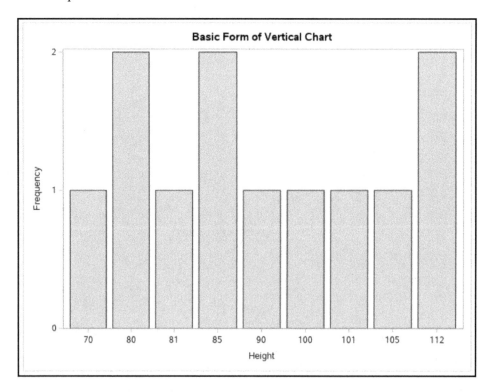

There are only three data points of Height, which have a frequency of 2.

Up until now, we have only explored a few of the data axis options. Let's experiment a bit with our basic vertical chart and try out some charting options:

```
Proc SGPLOT Data=Class;
  VBar Height / Dataskin=Gloss Stat=PCT;
   Title 'Vertical Gloss Chart with PCT';
Run;
```

This will result in the following chart as output:

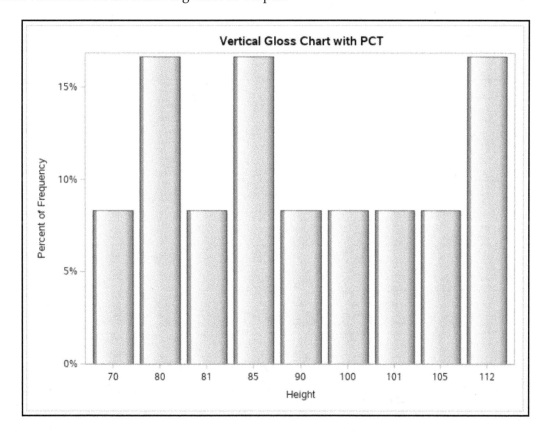

The other options for Dataskin are crisp, matte, pressed, and sheen. The default option is none. The other choices for Stat are mean, medium, and sum. The default option is freq.

Another aspect of the chart that you may want to control is the spacing between the bars. You can do this using the Barwidth option:

```
Proc SGPLOT Data=Class;
   VBar Height / Dataskin=Sheen Barwidth=0.5;
   Title 'Vertical Sheen Chart with Spread Out Bars';
Run;
```

Along with adding the Barwidth option, we have also changed Dataskin for the following chart:

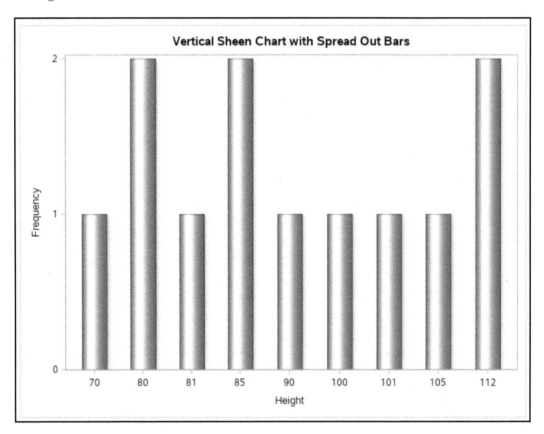

The `Barwidth` option can range from between 0 to 1. The higher the value, the wider the space between the bars. `Barwidth` achieves this by changing the width of the bar.

At some stage, you will want to call out the data points on your chart. The `Datalabel` option will come in handy for this:

```
Proc SGPLOT Data=Class;
  VBar Height / Datalabel Datalabelattrs=(family='Albany AMT'
  size=10pt color=red);
  Title 'Chart with Datalabel';
Run;
```

This will result in the following chart as the output:

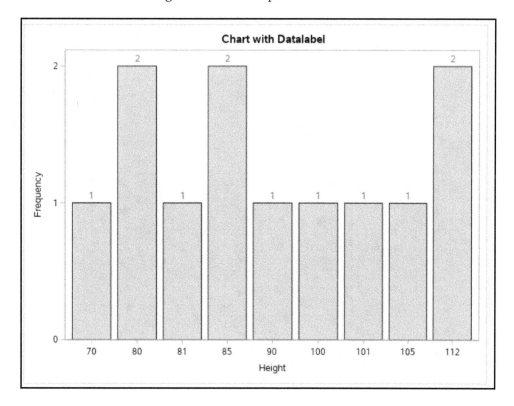

The data label's font, size, and color have also been specified.

Using the `Fillattrs` option, we have specified the fill color of the vertical bar. `Filltype` has two options—solid and gradient. Solid is the default option you have been seeing in almost all of the charts in this book:

```
Proc SGPLOT Data=Class;
   VBar Height / Datalabel Datalabelattrs=(Family='Albany AMT'
   Size=10pt Color=Red)
   Fillattrs=(Color=Blue) Filltype=Gradient;
   Title 'Chart with Color and Gradient';
Run;
```

The following chart is the resultant output:

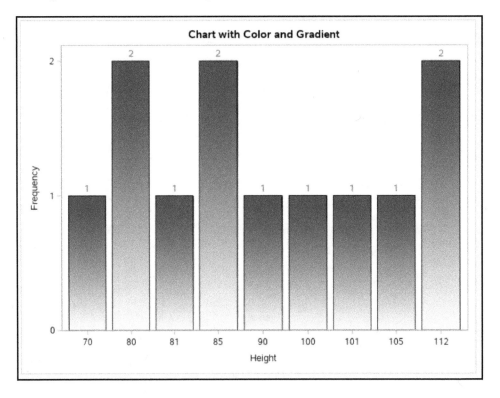

The `SGPLOT` procedure can also be utilized for building charts based on the statistical measures that are generated as part of the output of various procedures. We will use the `Means` procedure to calculate the mean and other statistical measures:

```
Proc Means Data=Class Alpha=.05 CLM Mean Std NoPrint;
   Class Year;
   Var Height;
   Output Out=ClassMean UCLM=UCLM LCLM=LCLM Mean=Mean;
Run;
```

The output dataset from `Proc Means` will be used in the `SGPLOT` statement to produce vertical bar charts. The `Vbar` statement that was used in the previous examples to produce charts has been replaced with the `Vbarparm` statement:

```
Proc SGPLOT Data=ClassMean;
   Vbarparm Category=Year Response=Mean / Datalabel;
Run;
```

This will generate the following chart as output:

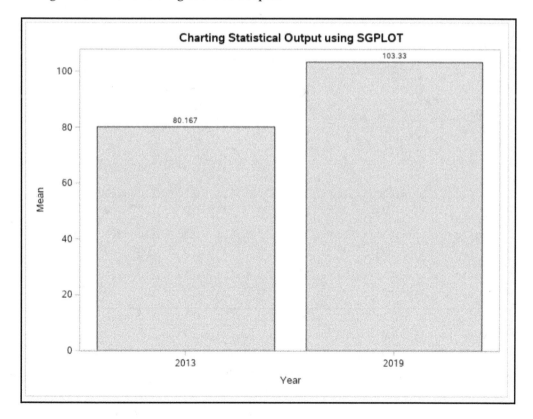

Here, the `Datalabel` option has been included to print the mean values.

The `SGPLOT` procedure has some built-in statistical measures, which we showcased by using the mean values while creating `Age` with a response chart. Similar output to the preceding chart can be created using the `SGPLOT` procedure without running `Proc Means`:

```
Proc SGPLOT Data=Class;
   Vbar Year / Response = Height Stat=Mean Limits=Upper
   Datalabel;
   Title 'Alternative Method for Charting Statistical Output';
Run;
```

This produces the following output:

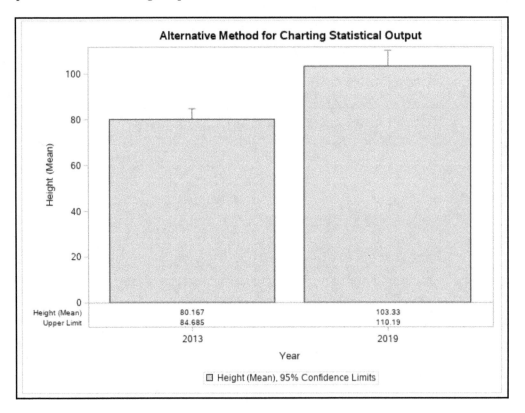

Similar to the chart-generated `SGPLOT`, we have two vertical bars representing the years and showcasing the mean with `Datalabels`. Additionally, we have been able to calculate the upper limit and showcase it in the preceding chart.

We can also include multiple statistical measures in the same chart using the following code:

```
Proc SGPLOT Data=Dealership;
  Vbar Day / Response = Units Stat=Mean Fillattrs=(Color=Blue)
  Datalabel Datalabelpos=Data;
  Vbar Day / Response = Units Stat=Median Datalabel
  Datalabelpos=Bottom Fillattrs=(Color=Red)
  Barwidth=0.5 Transparency=0.7;
  YAxis Display=None;
  Title 'Overlaying Vertical Bars';
Run;
```

This produces the following chart:

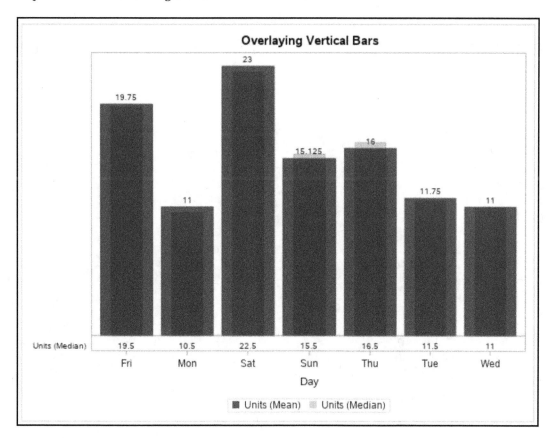

To overlay the two charts, we have added the transparency option. It has a value between 0 and 1—the higher the value, the more transparent the bar. At the bottom of the bar, we have the median values while, at the top of the bar, we have the mean. Using this method, you can overlay multiple charts. You may experience some practical problems if the scale of the measures you are trying to plot is quite different.

Bar charts are usually produced for grouped variables. We will use the following code to group various age groups and the heights of students within those:

```
Proc SGPLOT Data=Class;
   Vbar Age / Group= Height Stat=Percent Datalabel Datalabelpos=Data;
   Title 'Vertical Grouped Bar Chart';
Run;
```

We get the following chart as output:

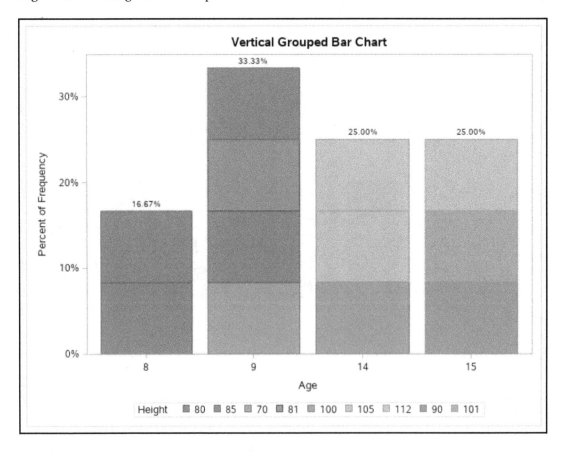

The code for horizontal bars is almost similar, except that you will have to use the `Hbar` statement instead of `Vbar`:

```
Proc SGPLOT Data=Class;
  Hbar Height;
   Title 'Horizontal Bar Chart';
Run;
```

The following chart is produced:

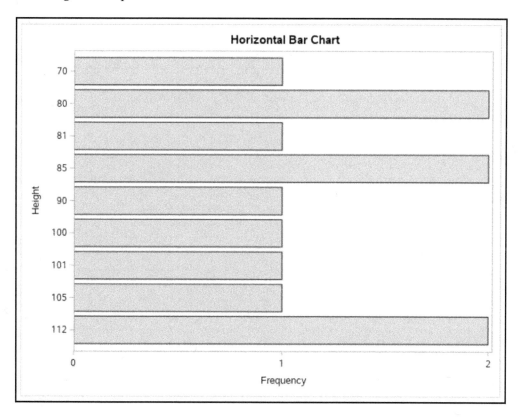

Scatter charts

These diagrams are used to establish whether there is a correlation between the variables that have been plotted.

If you remember the `Cost_Living` dataset well, you may recollect the `Other` variable. We don't know how it contributes to the cost of living index that's calculated for each city. Let's try and understand the relationship between `Other` and `Index` using scatter plots. As always, we will start with the simple form of scatter plot before we try and mix things up with the options and functionalities that the scatter plot offers us in the SAS environment:

```
Title "Index and Other Relationship";
  Proc SGPLOT Data=Cost_Living;
  Scatter X=Index Y=Other;
Run;
```

This will result in the following chart as the output:

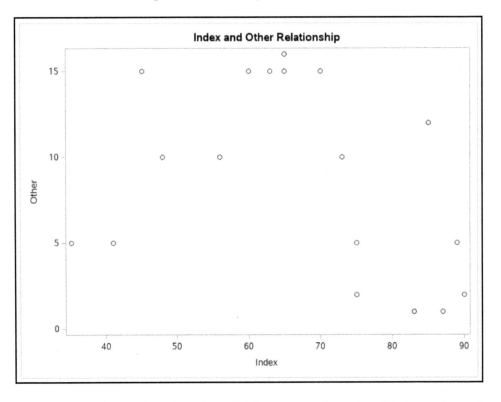

From the dataset, we know that the value of **Other** ranges from 1 to 16. Apart from the value of 5 for **Other** when the **Index** value is 35 and 41 for Johannesburg and Manila, respectively, the only other time the value of **Other** is 5 or lower is when the **Index** value is 75 or higher. In fact, the highest value of **Other** is observed when the **Index** value is in the medium range of values that the **Index** variable has to offer.

This seems to suggest that, at the lower and higher end of the **Index** value, individuals are spending on the **Other** category.

The other available constituents of `Index` can explain where most of the spending is happening. However, in the medium range of `Index` values, the `Other` variable is larger, which means that the variables available aren't capturing where individuals are spending money on the cost of living. We can come up with these and many more hypotheses by merely creating a scatter plot of two variables from a dataset that we have used extensively in this book.

We will now do a grouping of two variables to try and see how the variables relate to the group. We will use the following code to do this:

```
Title "Index and Prev Year Index's relationship with Housing";
  Proc SGPLOT Data=Cost_Living;
  Scatter X=Index Y=Prev_Yr_Index / Group=Housing
  Markerattrs=(Symbol=Circlefilled Size=3.5mm);
Run;
```

This results in the following chart as the output:

This is an interesting chart as it shows an almost linear relationship between Index and the previous years' Index values. In fact, all of the variables in the dataset, when used as the grouping variable, show a similar relationship between Index and the previous years' Index. This could be because the cost of living index is a slow-moving data element and isn't expected to relatively increase or decrease compared to a host of other cities. However, we might not have observed a linear relationship in the preceding chart if Index had been compared across a decade or more.

We used the Circlefilled option to make the colors more prominent and used the Size option for the same reason. We ended up with bigger circles than the ones we created in a default size scenario.

We will revert to the Class dataset to create paneled scatter plot:

```
Title 'Scatter via SGPANEL';
  Proc SGPANEL Data=Class;
  Panelby Year;
  Scatter X=Height Y=Weights;
Run;
```

The preceding code produces the following paneled scatter plots:

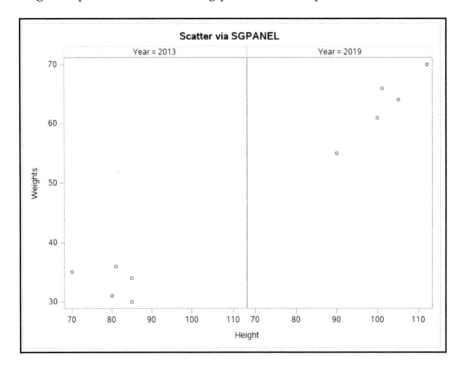

We used the SGPANEL procedure to produce a scatter plot. As you can see, a scatter plot can also be created outside of the SGPLOT procedure. There are various other SAS procedures that can produce scatter and other plots.

Suppose we want to find the standard error of the average price that's achieved on various days from the `Dealership` dataset. We are not worried about the average price of each car type or which team has sold it. Let's create datasets that calculate the average price per day, calculate the standard error of the average price per day, store the standard error in a macro variable, and then create a new dataset that has all of the required information to plot a standard error on a scatter graph:

```
Proc Sql;
   Create Table Day As Select Day, Avg(Avg_Price) as Avg_Price From
Dealership
      Group By 1;
Quit;

Proc Summary Data=Day;
   Var Avg_Price;
   Output Out=Day_Temp Stderr=Avg_Price_Stderr;
Run;

Proc Sql NoPrint;
   Select Avg_Price_Stderr Into: Avg_Price_Stderr From Day_Temp;
Quit;

Proc Sql;
   Create Table Day_Stderr As Select *, Avg_Price-&Avg_Price_Stderr as
Lower,
      Avg_Price+&Avg_Price_Stderr as Upper From Day;
Quit;

Title 'Scatter Std Error';

Proc SGPLOT Data=Day_Stderr;
   Scatter X=Day Y=Avg_Price / YErrorLower=Lower YErrorUpper=Upper;
Run;
```

This produces the following chart:

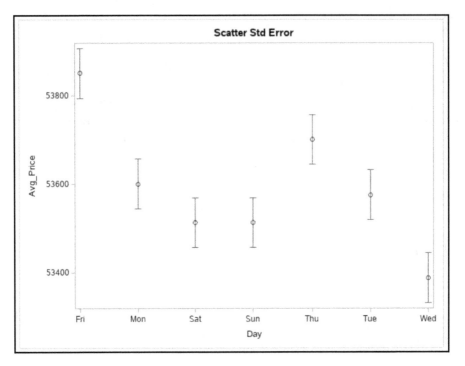

From the preceding chart, we can infer that the standard error of the average price does not vary significantly across the days. However, the average price does vary significantly.

Box plot

The **box and whiskers plot**, or **box plot**, it's as popularly called in SAS, is a plot of measurement organized in groups. The box plot displays the mean, quartiles, and minimum and maximum observations for a group. The benefit of a box plot is that it can display a variable's location and spread. It can showcase outliers and provide insight into the skewness of the data:

```
Title 'Basic Form of Box Plot';
  Proc SGPLOT Data=Class;
  VBox Height / Category=Year;
Run;
```

This produces the following chart:

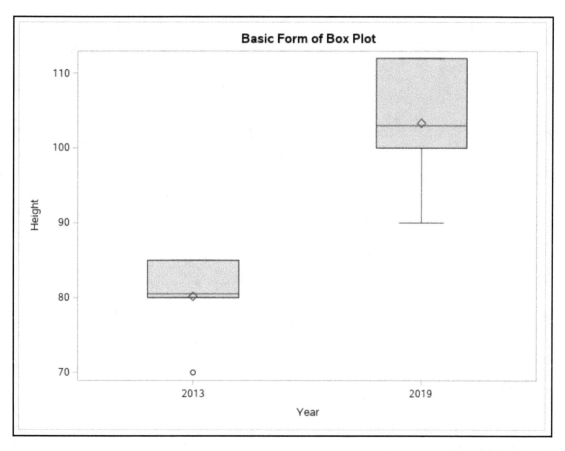

As you can see from the **Box Plot**, the year **2019** has more variance in the height of students than in **2013**.

We can also use the built-in **Box Plot** procedure as an alternative to the preceding use of the SGPLOT procedure:

```
Proc Boxplot Data=Class;
  Plot Height*Age;
    Inset Min Mean Max Stddev / Header='Height Statistics' POS=RM;
Run;
```

This produces the following chart, with statistics in the right-hand margin:

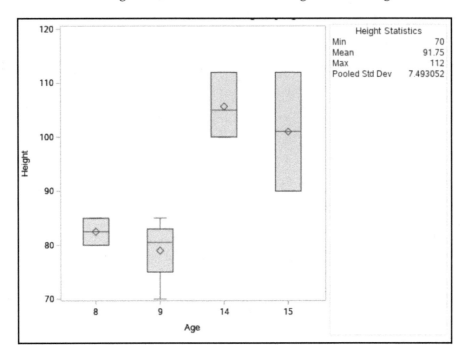

From the previous chart, we can infer that the **Height** in the same **Age** group varies more as the **Age** increases.

Summary

Throughout this book, we've explored various programs that help create, combine, reformat, transpose, and summarize datasets and perform other data-related actions. In this chapter, we learned about the role of data visualization. The various charts that were discussed in this chapter should help you narrate your data story. Plotting the same data in various charts should have highlighted the pros and cons of using charts by showcasing how the one-chart-fits-all data approach won't be the best option for your data.

In the next chapter, we will learn about the reporting and output delivery system.

Reporting and Output Delivery System

10

In this chapter, we will learn about sending our output to an external system and controlling what is sent and how the output is formatted in the file created in the external system. To do this, we will use the **Output Delivery System** (**ODS**). ODS helps to write SAS output to a particular destination. There are two types of destinations that are supported. The destinations and various formats within that are as follows:

- SAS formatted destinations. These include the following:
    - Document
    - Listing
    - Output
- Third-party formatted destinations. These include the following:
    - Excel
    - PowerPoint
    - EPub
    - HTML
    - **Rich Text Format** (**RTF**)
    - Printer family of destinations
    - Markup family of destinations

The default ODS destination we have seen throughout this book is Output. The internal data structure and the resultant output are similar in this case. The Document destination allows us to rearrange, restructure, and reformat the output data without rerunning the code. The Listing output helps to create the same look and feel of the output as supported in the previous versions of SAS.

In this chapter, we will focus on third-party destinations. Among these, the primary focus will be on Excel as this provides the greatest opportunity to reformat the data and rerun further analysis once the data is sent to Excel. We will introduce the Tabulate procedure in this chapter, and leverage our understanding of other procedures gained throughout this book.

The following topics will be covered in this chapter:

- Proc Tabulate
- Specifying ODS destinations
- Formatting ODS files
- ODS Excel

Proc Tabulate

In this book, we explored Proc Means, Summary, and Freq among a few of the available statistical summary functions in SAS. The Tabulate procedure incorporates many of the features of all of these. It is the ideal procedure if you want to publish some tables. It doesn't do anything new that the other procedures can't. It is fundamentally more adept at handling multiple variables and multiple levels of classes. It is good at computing multiple statistics and packaging them nicely so that you can publish your results.

Comparing multiple Proc Means and Proc Tabulates

Let's revisit a Proc Means example we looked at in `Chapter 4`, *Power of Statistics, Reporting, Transforming Procedures, and Functions*. We used the following `Customer_X` dataset:

ID	Class	Height	Weight	Football	Basketball	Hockey
1	A	Over5.7	Above50	1	0	1
2	A	Over5.7	Above50	1	1	0
3	B	Over5.7	Below50	1	1	.
4	B	Under5.7	Below50	1	1	1
5	A	Over5.7	Below50	1	1	1
6	A	Over5.7	Above50	1	.	1

Let's produce the means of the height of the `Basketball` players for the overall dataset and the two classes of the `Height` variable.

To produce the overall mean, we will run the following code:

```
Proc Means Data=Customer_X;
  Var Basketball;
Run;
```

This will give the following table as output:

The MEANS Procedure				
Analysis Variable : Basketball				
N	**Mean**	**Std Dev**	**Minimum**	**Maximum**
5	0.8000000	0.4472136	0	1.0000000

To generate the means for the various classes in the `Height` variable, we will run the following code:

```
Proc Means Data=Customer_X;
  Class Height;
  Var Basketball;
Run;
```

This produces the following output:

The MEANS Procedure						
Analysis Variable : Basketball						
Height	**N Obs**	**N**	**Mean**	**Std Dev**	**Minimum**	**Maximum**
Over5.7	5	4	0.7500000	0.5000000	0	1.0000000
Under5.7	1	1	1.0000000	.	1.0000000	1.0000000

To compare this to `Proc Tabulate`, we will run the following code:

```
Proc Tabulate Data=Customer_X;
  Class Height;
  Var Basketball;
  Table Basketball*Mean, Height All;
Run;
```

We get the following `Proc Tabulate` overall `Mean` and by `Height` output:

		Height		All
		Over5.7	Under5.7	
Basketball	**Mean**	0.75	1.00	0.80

The output of `Proc Tabulate` in the preceding output is available in a single table, whereas the same information is spread across the two outputs when we use the Means procedure. Hence, from a data publishing perspective, Proc Tabulate should be the default choice when publishing statistics.

The general syntax of `Proc Tabulate` is as follows:

```
Proc Tabulate <option(s)>;
```

The various statements supported within Proc Tabulate are as follows:

- `By`
- `Class`
- `Classlev`
- `Freq`
- `Keylabel`
- `Keyword`
- `Table`
- `Var`
- `Weight`

We will explore some of these statements in this chapter.

Remember that all variables mentioned in the `Table` statement should be included in the `Var` or `Class` statement. The order of the variables in the output will be determined by their order in the statement.

The asterisk (*) was used in the code to produce the `Proc Tabulate` overall `Mean` and by `Height` as, by doing so, we could add a statistic to the output. The other reason to use it in Proc Tabulate can be for adding a classification variable. The `ALL` option was used to calculate column totals. It can also be used to calculate row totals.

Multiple tables using Proc Tabulate

In the previous example, we saw how the output in multiple tables produced by Proc Means was produced in a single table using Proc Tabulate. There is a feature of Proc Tabulate that allows you to create multiple tables using the same program, albeit with the use of multiple `Table` statements in the same program:

```
Proc Tabulate Data=Dealership;
   Var Units;
   Class Car Team Day;
   Table Units;
   Table Car Team Day;
Run;
```

This produces the following table with the sum of units sold:

Units
Sum
583.00

It also produces a summary of units sold by car type and selling team, and the units sold on each day:

Car		Team				Day						
Alpha	Omega	A1	A2	A3	A4	Fri	Mon	Sat	Sun	Thu	Tue	Wed
N	N	N	N	N	N	N	N	N	N	N	N	N
18	18	9	9	9	9	4	4	8	8	4	4	4

Choosing the statistics

For generating the Proc Means of basketball by `Height`, we specified that we wanted the `Means` statistic but still got the number of observations and the minimum, maximum, and standard deviation in the output. In the rest of the Proc Tabulate outputs discussed till now, we have seen the reporting of the number of observations and the sum of observations. We can use the following code to specify the statistics required:

```
Proc Tabulate Data=Dealership;
   Class Car;
   Var Avg_Price;
   Table Avg_Price*Car*(Sum Mean StdDev);
Run;
```

This produces the following output:

Avg_Price					
Car					
Alpha			Omega		
Sum	Mean	StdDev	Sum	Mean	StdDev
708800.00	39377.78	232.14	1219850.00	67769.44	431.53

The parentheses used in the code to produce this output was done to simplify understanding the code as it clubbed variables with the same data treatment.

The alternative to the parentheses would use the following code:

```
Proc Tabulate Data=Dealership;
   Class Car;
   Var Avg_Price;
   Table Avg_Price*Car*Sum Avg_Price*Car*Mean
   Avg_Price*Car*StdDev;
Run;
```

The `Table` statement looks more complicated in this instance compared to the statement that produced the preceding output. The complicated version of the code produces the following output:

Avg_Price		Avg_Price		Avg_Price	
Car		Car		Car	
Alpha	Omega	Alpha	Omega	Alpha	Omega
Sum	Sum	Mean	Mean	StdDev	StdDev
708800.00	1219850.00	39377.78	67769.44	232.14	431.53

Formatting the output

When the output is produced, the headers have the name of the variables. This can be modified to make the tables self-explanatory. We can add the narration in the `Table` statement in the `Tabulate` procedure:

```
Proc Tabulate Data=Customer_X;
   Class Height;
   Var Basketball;
```

```
    Table Basketball="No. of Students Playing Basketball"*
    Sum="Total No. of Students"*Height;
  Run;
```

This produces the following output:

No. of Students Playing Basketball	
Total No. of Students	
Height	
Over5.7	Under5.7
3.00	1.00

Without formatting, the output would have been as follows:

Basketball	
Sum	
Height	
Over5.7	Under5.7
3.00	1.00

Clearly, the interpretation of the formatted output is more intuitive than the unformatted output in the preceding screenshot.

Two-dimensional output

Till now, we have focused on one-dimensional Proc Tabulate output. We will use a new dataset to look at the two-dimensional output:

```
  Data Sales;
    Input Country $7. Segment $11. Type $ Product $ Amt;
    Datalines;
  US        Retail       Software A 23
  US        Retail       Software B 11
  US        Retail       Hardware A 8
```

```
US        Retail      Hardware B 10
US        Commercial Software A 45
US        Commercial Software B 46
US        Commercial Hardware A 4
US        Commercial Hardware B 11
Germany Retail        Software A 12
Germany Retail        Software B 15
Germany Commercial Software A 55
Germany Commercial Software B 67
Germany Commercial Hardware A 23
Germany Commercial Hardware B 25
;
```

The code generates the following table, which is the `Sales` dataset:

Country	Segment	Type	Product	Amt
US	Retail	Software	A	23
US	Retail	Software	B	11
US	Retail	Hardware	A	8
US	Retail	Hardware	B	10
US	Commercial	Software	A	45
US	Commercial	Software	B	46
US	Commercial	Hardware	A	4
US	Commercial	Hardware	B	11
Germany	Retail	Software	A	12
Germany	Retail	Software	B	15
Germany	Commercial	Software	A	55
Germany	Commercial	Software	B	67
Germany	Commercial	Hardware	A	23
Germany	Commercial	Hardware	B	25

We want the output that calculates the mean of hardware and software sales amount for each country, segment, and product. Let's use the following code to do so:

```
Proc Tabulate Data=Sales;
   Class Country Segment Type Product;
   Var Amt;
   Table Country*Segment*Product,Amt*Type*Mean;
Run;
```

This gives the following table as the output:

Country	Segment	Product	Amt Hardware Mean	Amt Software Mean
			Amt	
			Type	
			Hardware	Software
			Mean	Mean
Germany	Commercial	A	23.00	55.00
		B	25.00	67.00
	Retail	A	.	12.00
		B	.	15.00
US	Commercial	A	4.00	45.00
		B	11.00	46.00
	Retail	A	8.00	23.00
		B	10.00	11.00

This is a two-dimensional Proc Tabulate output.

Specifying the ODS destination

To specify the ODS destination for Excel, the syntax is as follows:

```
ODS EXCEL FILE="filename.xlsx";
```

We will use the Class dataset used frequently in this book. The filename specified is the same as the dataset. The following destination relates to the project folder I have created as part of my SAS University Edition session. This needs to be customized as per the SAS version you are using. It might be difficult to produce this output in a client-server location unless you know the destination where you are allowed to save and retrieve files:

```
ODS Excel File = '/folders/myfolders/Class.xlsx';
Proc Print Data=Class;
Run;
ODS Excel Close;
```

In the preceding code, we have opened a new ODS destination and closed it at the end of the `Proc Print` procedure as we don't want the rest of the session output to go to the Excel location. The code produces the following output:

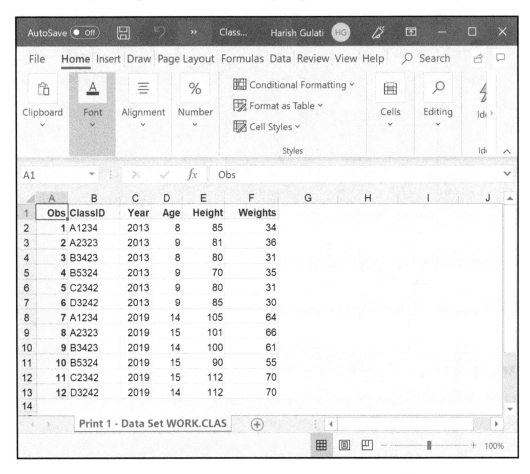

Alternatively, you can use the following code to produce the output in PDF:

```
ODS PDF File = '/folders/myfolders/Class.pdf';
Proc Print Data=Class;
Run;
ODS PDF Close;
```

Now that we will have seen how to specify the ODS destination, we will learn how to format the ODS file.

Formatting ODS files

We will now look into the various techniques of formatting the ODS files.

Multiple sheets

If we have *n* number of output files, we don't need to produce *n* number of destination files. For example, you can use the following code to put three of our datasets into the same Excel file. Each dataset will be written to a different sheet, and the name of the sheet is specified in the code:

```
ODS Excel File='/folders/myfolders/Datasets.xlsx'
Options(Sheet_Name="Class");

Proc Print Data=Class;
Run;

ODS Excel Options(Sheet_Name="Customer_X");
;

Proc Print Data=Customer_X;
Run;

ODS Excel Options(Sheet_Name="Dealership");
;

Proc Print Data=Dealership;
Run;

ODS Excel Close;
```

The code produces the following file in the specified location:

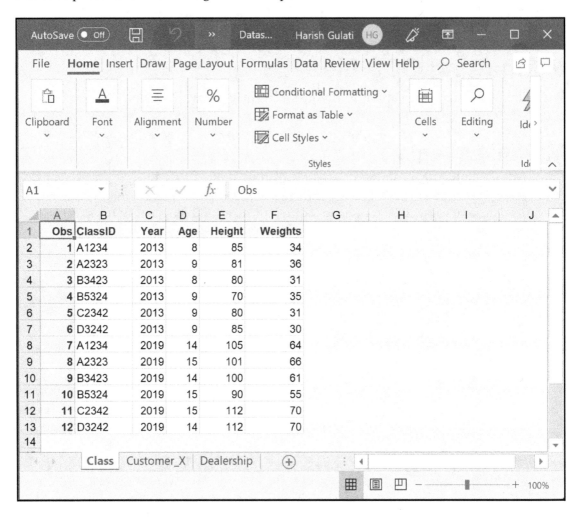

In this example, we had three datasets that were populated in individual sheets in an Excel file.

A single dataset can also have its information populated in multiple sheets using the BY group:

```
ODS Excel File = '/folders/myfolders/Multiple_Datasets.xlsx';
Proc Report Data=Class;
   By Year;
Run;
ODS Excel Close;
```

This produces multiple sheets in the Excel file, `Multiple_Datasets.xlsx`:

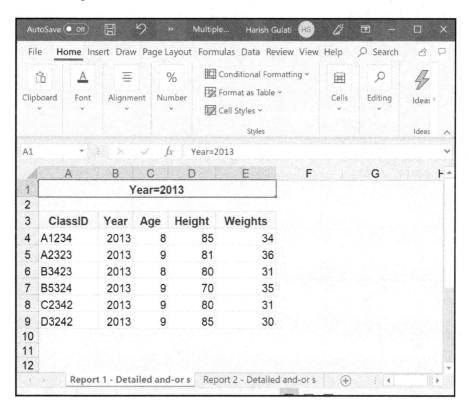

The second sheet produced contains the data of 2019, which is the second level of the BY group in the class dataset:

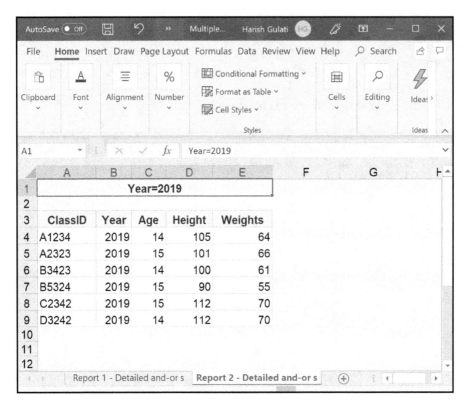

Applying filters

To apply filters to the destination file, use the following code:

```
ODS Tagsets.Excelxp File='/folders/myfolders/Filters.xls' Style=statistical
   Options (Autofilter='all');
Proc Print Data=Class;
Run;
Ods Tagsets.Excelxp Close;
```

This produces the following file:

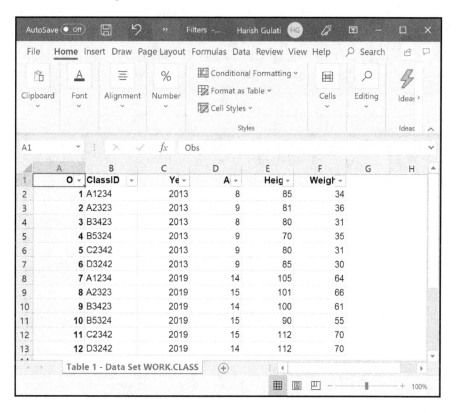

Controlling the print options

It is possible to control your print options using the SAS code. This ensures that your output is not done uncontrollably. We will run the following code and look at how the options used in the code related to the print options:

```
ODS Excel File = '/folders/myfolders/Print_Dealership.xlsx' Options
(BlackandWhite='Yes' Center_Horizontal='Yes'
Center_Vertical='Yes' Draftquality='On');
  Proc Print Data=Dealership;
Run;
ODS Excel Close;
```

This code will produce the `Print_Dealership` file in the output location we specified. When you select **File** and then **Print** from the MS Excel menu, you can see that the data is aligned at the center of the page.

Further, when you click on **Page Setup** and look at the **Sheet** tab, you can observe that the **Black and white** and **Draft quality** options have been pre-selected due to the code that has been run. Please see the following screenshot for details:

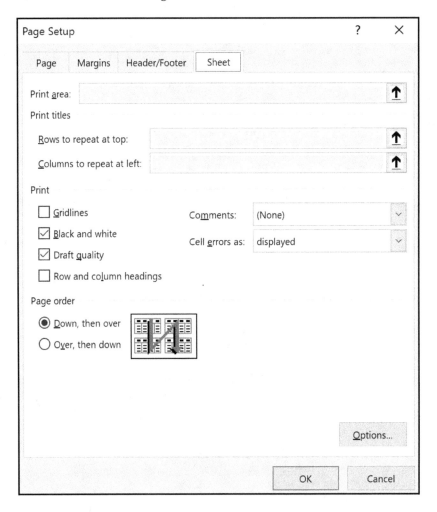

You can even specify whether it's portrait or landscape view, the print area, row-column headings, and the row height, among various other things.

One thing that you must look out for when trying the following code is that the Excel file won't give away all of the settings you have requested until you click on the **File | Print** option, and then look for them. The Excel file would have appeared the same with or without the settings we have requested, except for the effect of the heading row height change:

```
ODS Excel File = '/folders/myfolders/Print_More_Options.xlsx' Options
(Print_Area='B,2,G,11' RowColHeadings='Yes' Row_Heights='40'
Orientation='Landscape');
   Proc Print Data=Dealership;
Run;
ODS Excel Close;
```

This produces the following file:

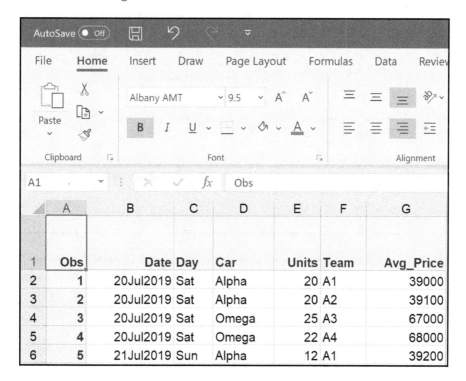

This is a partial view of the **Print More Options** file.

The landscape option can be selected and viewed in the **File | Print** options in the Excel file. Furthermore, you can see that the **Print Active Sheets** option is selected and only the 10 rows of data are available to print along with the header, as per our request via the `Print_Area` option:

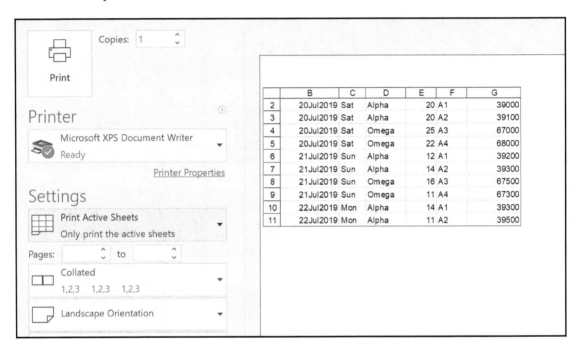

It can be seen from the Excel sheet how many pages would be required to print the file. This can be achieved by using the `Rowbreaks_Count` option. You will need to ensure that the **Page Break** option is turned on in the Excel file before you can view the preselected option via the following code:

```
ODS Excel File='/folders/myfolders/Row_Break.xlsx'
  Options (Rowbreaks_Count='2');

Proc Print Data=Class;
Run;
ODS Excel Close;
```

This produces the following file:

	A	B	C	D	E	F
1	Obs	ClassID	Year	Age	Height	Weights
2	1	A1234	2013	8	85	34
3	2	A2323	2013	9	81	36
4	3	B3423	2013	8	80	31
5	4	B5324	2013	9	70	35
6	5	C2342	2013	9	80	31
7	6	D3242	2013	9	85	30
8	7	A1234	2019	14	105	64
9	8	A2323	2019	15	101	66
10	9	B3423	2019	14	100	61
11	10	B5324	2019	15	90	55
12	11	C2342	2019	15	112	70
13	12	D3242	2019	14	112	70
14						

The Excel sheet shows the page break information.

Changing the default cells

For producing the desired print settings in the partial view of the Print More Options output and the landscape view data, we have already seen the use of cell selection in the options. The output cells can also be selected to control where the data is pasted in the Excel file. Let's use the following example to paste the data in row 2 and column 3:

```
ODS Excel File = '/folders/myfolders/Start_Pos.xlsx' Options
(Start_at='2,3');
Proc Print Data=Class;
Run;
ODS Excel Close;
```

This produces the following output:

	A	B	C	D	E	F	G
1							
2							
3		Obs	ClassID	Year	Age	Height	Weights
4		1	A1234	2013	8	85	34
5		2	A2323	2013	9	81	36
6		3	B3423	2013	8	80	31
7		4	B5324	2013	9	70	35
8		5	C2342	2013	9	80	31
9		6	D3242	2013	9	85	30
10		7	A1234	2019	14	105	64
11		8	A2323	2019	15	101	66
12		9	B3423	2019	14	100	61
13		10	B5324	2019	15	90	55
14		11	C2342	2019	15	112	70
15		12	D3242	2019	14	112	70

We see the B3 cell is the starting position in the output.

ODS Excel charts

It's not just data that we can export to Excel and other destinations. We can also export charts. By default, when you run multiple procedures (be it for producing data or chart output) in the same ODS statement, multiple sheets are created in Excel. Use the `Sheet_interval` option to ensure that you get the output in the same sheet if needed:

```
ODS Excel File = '/folders/myfolders/Chart_Graph_Same_Page.xlsx'
  Options (Sheet_interval='None');
Proc Means Data=Class;
  Var Height;
Run;
Proc SGPLOT Data = Class;
  Histogram Height;
  Title 'Height of children in class across years';
Run;
ODS Excel Close;
```

This produces the following output:

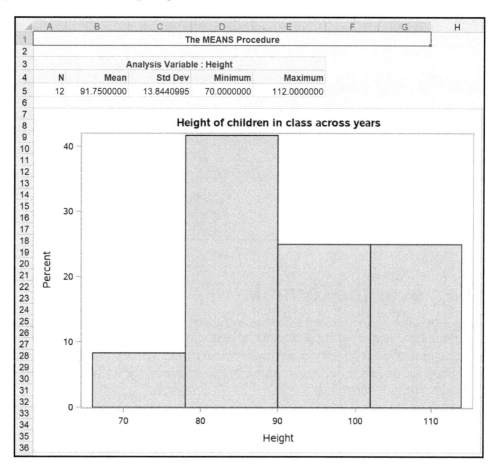

Color-coding the output

We can color-code the output based on certain values. This is also known as traffic lighting the data you generate. Remember that, in the following code, we could have used the "low-80" syntax for specifying the least height to have the color red. However, doing so would have also ensured that the `Age` variable would have been color-coded:

```
Proc Format;
Value Format_Height
80-90 ='Red'
91-110='Yellow'
111-high='Green';
```

```
ODS Excel File = '/folders/myfolders/Proc_Format.xlsx';
Proc Print Data=Class NoObs;
Var ClassID Age Height / Style=[Backgroundcolor=Format_Height.];
Where Year=2019;
Run;
ODS Excel Close;
```

This produces the following output:

	A	B	C
1	ClassID	Age	Height
2	A1234	14	105
3	A2323	15	101
4	B3423	14	100
5	B5324	15	90
6	C2342	15	112
7	D3242	14	112

Copying over the formula

The biggest drawback of ODS is that you may never see how the data has been generated. Think about the Dealership data. It doesn't have the total revenue column. Yes, we have generated the total revenue in previous codes. However, can we send the formula of total revenue to Excel so that, for a derived column, the user can see the underlying formula? Let's try it using the following code:

```
ODS Excel File = '/folders/myfolders/Formula.xlsx';
Options Obs=5;
Proc Report Data=Dealership;
  Column Date Car Units Avg_Price Total_Revenue;
  Define Units / Display;
  Define Avg_Price / Display;
  Define Total_Revenue / "Total_Revenue" Computed
  Format=Dollar10.2
    Style={TagAttr="Formula:(RC[-2]*RC[-1])"};
  Compute Total_Revenue;
    Total_Revenue=Units*Avg_Price;
  Endcomp;
Run;
ODS Excel Close;
```

The following output is produced:

E2		▼	:	×	✓	fx	=(C2*D2)	

	A	B	C	D	E
1	Date	Car	Units	Avg_Price	Total_Revenue
2	20Jul2019	Alpha	20	39000	$780,000.00
3	20Jul2019	Alpha	20	39100	$782,000.00
4	20Jul2019	Omega	25	67000	$1,675,000.00
5	20Jul2019	Omega	22	68000	$1,496,000.00
6	21Jul2019	Alpha	12	39200	$470,400.00

As we can see, the column statement declares the variables that need to be included in the output. You may have noticed that the newly computed variable has been mentioned in this statement. Despite including `Units` and `Avg_Price` in the column statement, if you don't mention them in the `Define` statement, they will remain uninitialized to be used in the `compute` statement. Without the `Define` statement, they will be included in the output dataset but not get initialized to compute the new variable. The `RC[-2]` and `RC[-1]` commands mean that the first variable for the formula is two spaces to the left, and the second variable for the formula is one space to the left.

Summary

In this chapter, which happens to be the last in this book, we focused on learning about the Tabulate procedure, which is a combination of a few other statistical procedures, and we learned about ODS. We looked at generating one- and two-dimensional tables in Proc Tabulate that incorporated various types of statistical measures. We compared its output with the `Means` procedure to showcase the benefits of using Proc Tabulate. In the second half of this chapter, we turned our attention to discovering all of the destinations that ODS can help share our SAS data with. Using Excel as the key focus, we learned about various programming options that can enhance the quality and presentation of the data that can be shared with Excel via the SAS programs.

Other Books You May Enjoy

If you enjoyed this book, you may be interested in these other books by Packt:

SAS for Finance
Harish Gulati

ISBN: 9781788624565

- Leverage the power of SAS to analyze financial data with ease
- Find hidden patterns in your data, predict future trends, and optimize risk management
- Learn why leading banks and financial institutions rely on SAS for financial analysis

Big Data Analytics with SAS
David Pope

ISBN: 9781788290906

- Combine SAS with platforms such as Hadoop, SAP HANA, and Cloud Foundry-based platforms for efficient Big Data analytics
- Learn how to use the web browser-based SAS Studio and iPython Jupyter Notebook interfaces with SAS
- Practical, real-world examples on predictive modeling, forecasting, optimizing and reporting your Big Data analysis with SAS

Leave a review - let other readers know what you think

Please share your thoughts on this book with others by leaving a review on the site that you bought it from. If you purchased the book from Amazon, please leave us an honest review on this book's Amazon page. This is vital so that other potential readers can see and use your unbiased opinion to make purchasing decisions, we can understand what our customers think about our products, and our authors can see your feedback on the title that they have worked with Packt to create. It will only take a few minutes of your time, but is valuable to other potential customers, our authors, and Packt. Thank you!

Index

about 55, 56, 57, 58, 60
  reference link 59
scatter charts 291, 292, 293, 294, 295, 296
SCL Compiler 163
SET statement 97
SQL views
  syntax 243
statistical tests 131, 132, 133, 134, 135
string identification
  about 55
  find function 62, 63, 64
  index function 60, 61
  indexc function 60, 61
  indexw function 60, 61
  scan function 55, 56, 57, 58, 60
strip function 64
Structured Query Language (SQL) 16
subsetting 233, 234, 235
summarizing 235, 237
sysdate 20

T

temporary variable
  leveraging 98, 99
Teradata naming convention
  in SAS 20, 21
tokenization 161

tokens, type
  literals 161
  names 161
  numbers 161
  special 161
trim function 64
two-tailed test 135

U

unique values 119
UNIVARIATE Procedure table 140
UpCase 52, 53
updating 91

V

validation 145
variable length 48, 49, 50, 51, 94, 96, 97
vertical bar charts 282, 284, 285, 286, 287, 288,
  289, 290, 291

W

WHERE statements 32, 33, 34
WHICHC function 79, 80
WHICHN function 79, 80
Word Scanner 163, 174
Work library 17